设计是一场心灵的修行

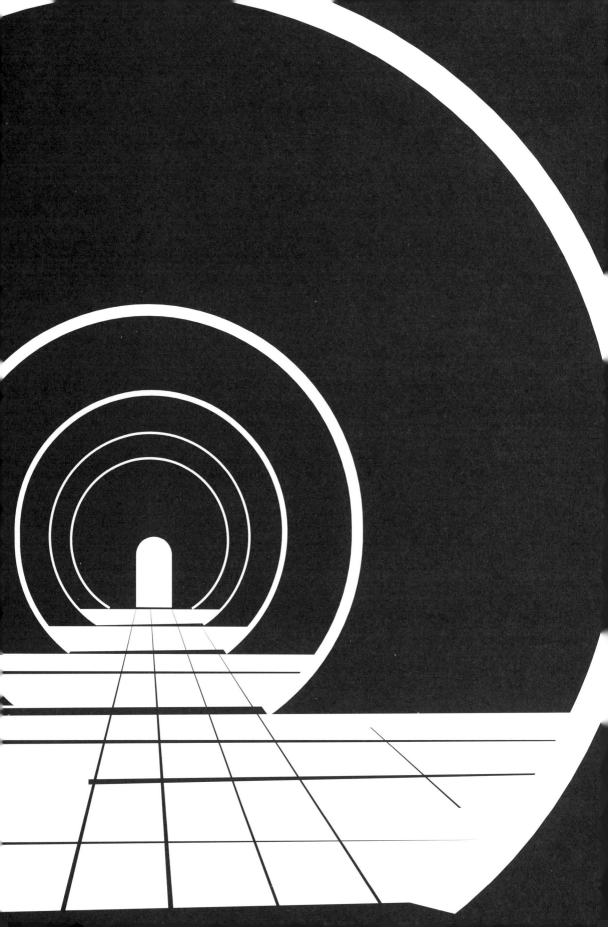

Inside Observation Outside Architecture

内观外筑

周湘华｜著

中国建筑工业出版社

序 I
Preface I

从 1992 年参与第一个设计开始，湘华在建筑领域深耕多年。这些年里，他"身处湘楚"不断"绽放芳华"，主持了大量的项目，可谓是作品等身，而本书就是他对诸多作品遴选之后的一次回溯与分享，当然也是对过往成绩的一种肯定。

从本书可以看出，湘华经手项目的种类和数量广博丰盈，可以说是一位全能选手。可不管建筑是什么规模，什么定位，他总是以一种专注和执着的态度投身其中，最后的成品不晦涩、不猎奇，雅致从容，细腻耐看，自有一种气质。这种气质是要在体察他建筑本体及其场域之后方能品味出来。场域一词源自社会学家布尔迪厄，它被定义为"在各种位置之间存在的客观关系的一个网络或一个构型"。从场域的视角来看，湘华的建筑形式就在所处区域特有的场地地形、地域气候、社会文化、人群行为、资本权力的要求、造价控制下适宜工程技术以及他个人的审美旨趣所构成的复杂网络中被赋形，被定义，被呈现，被传播，作品气质也自内而外弥散出来。浔龙河生态示范点公共配套项目设计就可看做一个典型。他手绘的要点网络图就直接将建筑的场域要素展现出来，包括项目所处时代特征、两型城市政策背景、资本与政府还有市场的利益共享关系、都市农业与康养医疗内容策划、单体建筑等所体现的"仁、义、礼、智、信"文化属性等。该图绘的表达方式也很有巧思，他以浔龙河的线性形态作为核心，由它发散关联到各个抽象要素，而这些关系线仿若河流的支流一般。抽象的关系和具体的环境形式、文字与图形相互交织融入，这是建筑师才能绘出的图解，让人击节回味。此处仅仅举一项，书中这样的例子比比皆是。从中我们可以看出他对环境鞭辟入里的观察和思索，这也为其创作提供了锚点。

上述内容是"外筑"的表达，湘华的探赜未至于此，这本作品集还把他更深层次的思考即"内观"展示出来。如果引入语言学相关观念，建筑外在的空间形式关系是表层结构，前文讨论的建筑场域是浅层结构，而"内观"则为深层结构。二者是一种层层决定的关系。湘华在多年的工作实践中，没有被世俗杂务蒙蔽本心，持续不断地在思想、文化艺术、环境层面求索，逐渐构成了自身的设计哲学观，即深层结构。它源自本人对佛道儒教义的体悟，传统与现代艺术的启迪以及现象学层面的思考。当然，这些思想的形成并非一蹴而就，而是经年累月沉淀而成，这自然影响到创作本身，所以我们也可以看到其作品的一些演进脉络：早期一些住宅作品形式执念较深，略显拘谨稚嫩，有斧凿之痕；后期作品（如张家界武陵源伍仟民宿）探求建筑与风土之间的关系，则更为恣肆从容，挥洒自如。此外，湘华还将"内观"与"外筑"是如何弥合一起的途径毫无保留地讲述出来，这包括建筑与时空的关联，设计与修心的关系以及当下与未来的绵延三个部分。由此，本书也建构出一个完整的包含思想、创作、实施各个层面的自身设计语言体系，能为还在不断探索中的湖南本土创作者带来了一定启示。

如果我没记错，湘华是 1972 年出生，现在已经是"知天命"的年纪，然而在建筑师的年龄宇宙中，他还属于"而立之年"，创作的高峰期刚刚到来。该书的出版既是以往工作的一个总结与终点，也是接下来创作生涯的基础与起点，愿他初心不忘，不断砺炼自身，保持与这个时代的高频共振，产生更精彩的作品！

湖南大学设计研究院有限公司董事长
2022 年 9 月于湖南大学

序 II

Preface II

　　本书是以"内"为出发点，向"外"探索，寻求建筑存在的自逻辑。由感知到可量度的过程，既是自然观，也是某阶段的起承转合；地域影射文化，建筑则自带出生印记。本书反其道行之，从具体事例逐渐抽象，慢慢表达，慢慢推进，慢慢讲述"现代，是历史传承组成"的故事。

　　1992 年湘华从学校毕业在省建科院工作至今30 余年，他对每个项目都做到精益求精，方案会经过反复的推敲，从项目的开始设计到建成，他几乎把所有的时间和精力都投入到项目中去，可以说每个作品都是他的心血之作。不仅自己学习不辍，而且带领团队积极打造学习型团队，取得了丰厚的成果，此书就是他寻找建筑设计的原点，细细讲述每个案例的出生、成长和落地，同时也是建筑师本人人生的起承转合。路易斯·康曾表述，"我已经达到了这一时刻，我明白如何以自己的方式作自我表达。这也是对待建筑物的一种方式、态度"。我相信作者也是经历沉淀、思考过后，懂得了如何用他自己的方式表达自己对建筑的理解和态度。

　　初读此书，看到的是创作者对每个案例的用心和思考，体现出了地域化的设计思考模式；再读时，看到的是BIM 在建筑行业数字化转型中的作用，与作者相伴的人生历程和时代变迁。从湘华诸多作品可以看出，无论建筑技术如何发展，每个建筑都有其注定的文化印记，每位建筑师都有其坚守的初心及设计的原点。每个建筑都是建筑师与光、与材料、与地域、与人文对话后的结果，静谧而有力量。我认为，我会带着这样的想法和思考再读此书，或许能收获另一份意想不到的惊喜。

　　读懂建筑，不仅只是第一眼地感动，而是其本身带给你的力量。就像此书的章节顺序，当你认真解读每个建筑作品，由外观到建筑内部，都会带给你无限的力量。

曾任湖南省建筑科学院研究院院长、总建筑师等职
2022 年 8 月于紫金苑

目录

Contents

建筑设计需要感性的直觉与理性的逻辑分析相结合。感性的直觉可以通过手绘草图的形式呈现，理性的逻辑分析可以使用思维导图。在设计过程中，要处理好道术器用的关系：道即方向，术即方法，器即工具，用为目的。以道御术，以术选器，以器行用，最终达到知行合一的过程，我们要做的是明道、优术、利器、践行。生态绿色低碳的设计手法，跟"道法自然"的关系紧密。

合

转

项目中存在诸多不足，需要设计师不断学习与阅读、实践与反思。虽然不是所有的项目都能建成，但是建筑设计的魅力不仅限于项目完成时的满足感，更多的是在过程中的磨砺与修行，并将经验与收获运用于后续项目，以达到更好的完成度。在 BIM 还不成熟的阶段，与外国建筑师合作武陵山文化产业园，开始为筑而研，以用为筑，有所为而有所不为。

起初只是想将项目中勾画的思维导图和手绘草图进行整理。2016 年初步计划整理项目展示构架，2017 年开始汇总工作室十年项目资料，由于项目较多，投入时间有限，许多项目正在进行中，积累素材不足，反思也不够。因此 2018 年重新选择项目，拟定初稿。2019 年进一步扩充素材和深化内容，不断梳理和反思设计过程中的得与失。2020 年疫情期间，系统地读了一些书，整理出一些素材，尤其是传统文化方面的。2021 年……

项目场地踏勘是设计师和场所反复交流的过程，不停地走动，眺望四周，聆听当地的故事，通过手绘草图快速记录，回想场地带来的最初感受，勾画出思维导图，清晰地梳理出设计思路。感恩时代，让我能够成为一名建筑师。岁月积累是时间的沉淀，抑扬顿挫、起承转合是

Architectural design requires the combination of perceptual intuition and rational logical analysis. Perceptual intuition can be presented in the form of hand-drawn sketches, and rational logical analysis can use mind maps. In the design process, it is necessary to deal with the relationship between Taoist instruments and instruments: the Tao is the direction, the technique is the method, the instrument is the tool, and the use is the purpose. To master the way, to choose the tools, to use the tools, and finally to achieve the process of the unity of knowledge and action. In addition, the ecological, green and low-carbon design method is closely related to "the way of nature".

Combine

Change

There are many deficiencies in the project, which requires designers to continuously learn and read, practice and reflect. Although not all projects can be completed, the charm of architectural design is not only the satisfaction at the completion of the project, but also the tempering and practice in the process, and applying the experience and gains to subsequent projects to achieve a better degree of completion . In the immature stage of BIM, Wulingshan Cultural Industry Park cooperated with foreign architects and began to research for construction, use for construction, and do something but not to do something.

Rise

Inherit

At first, I just wanted to organize the mind maps and hand-drawn sketches sketched in the project. In 2016, the preliminary plan was to sort out the project materials. In 2017, the studio began to summarize the ten-year project materials. However, due to the large number of projects and the lack of investment time, many projects were in progress and the accumulation of materials was insufficient, and the reflection was not enough. In 2018, the project was selected and a first draft was drawn up. In 2019, we will continue to expand the materials and deepen the content, and constantly sort out and reflect on the gains and losses in the design process. During the epidemic in 2020, I systematically read some books and organized some materials, especially the study of traditional culture. 2021……

The site survey of the project is a process of repeated communication between the designer and the site, walking around constantly, looking around, listening to local stories, quickly recording through hand-painted sketches, recalling the initial feelings of the site, sketching a mind map, and clearly sorting out the design ideas . Be grateful for the times and be able to become an architect. The precipitation of time, the rhythm, the ups and downs, the ups and downs, the ups and downs are the rhythms of space, and the beauty exists at the intersection of things.

起

Rise

《尔雅》： 妃、合、会，对也

《说文解字》： 运也。从车专声。知恋切

沙中淘金

每个项目都是成长的足迹，

或许不够好，但也是思考的过程，

每个项目都像星尘之微粒，

经过水的淘洗，手的塑形，火的砺炼，光的呈现，

最终成为或实用，或美观的一件作品。

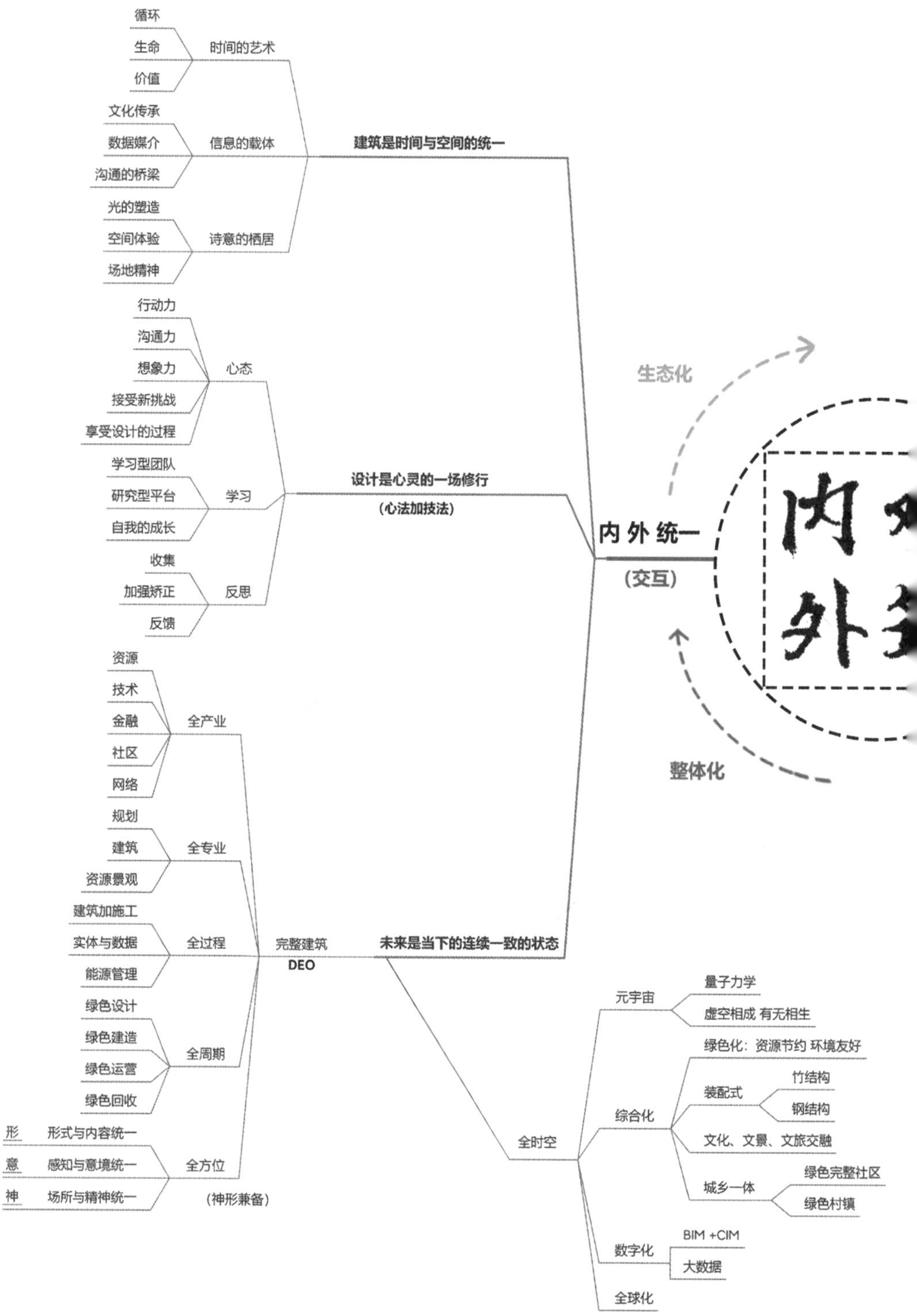

建筑是时间与空间的统一

- **时间的艺术**
  - 循环
  - 生命
  - 价值
- **信息的载体**
  - 文化传承
  - 数据媒介
  - 沟通的桥梁
- **诗意的栖居**
  - 光的塑造
  - 空间体验
  - 场地精神

设计是心灵的一场修行
（心法加技法）

- **心态**
  - 行动力
  - 沟通力
  - 想象力
  - 接受新挑战
  - 享受设计的过程
- **学习**
  - 学习型团队
  - 研究型平台
  - 自我的成长
- **反思**
  - 收集
  - 加强矫正
  - 反馈

完整建筑 DEO
未来是当下的连续一致的状态

- **全产业**
  - 资源
  - 技术
  - 金融
  - 社区
  - 网络
- **全专业**
  - 规划
  - 建筑
  - 资源景观
- **全过程**
  - 建筑加施工
  - 实体与数据
  - 能源管理
- **全周期**
  - 绿色设计
  - 绿色建造
  - 绿色运营
  - 绿色回收
- **全方位（神形兼备）**
  - 形 形式与内容统一
  - 意 感知与意境统一
  - 神 场所与精神统一

内 外 统一（交互）

生态化

整体化

内 外

全时空

- **元宇宙**
  - 量子力学
  - 虚空相成 有无相生
- **综合化**
  - 绿色化：资源节约 环境友好
  - 装配式
    - 竹结构
    - 钢结构
  - 文化、文景、文旅交融
  - 城乡一体
    - 绿色完整社区
    - 绿色村镇
- **数字化**
  - BIM +CIM
  - 大数据
- **全球化**

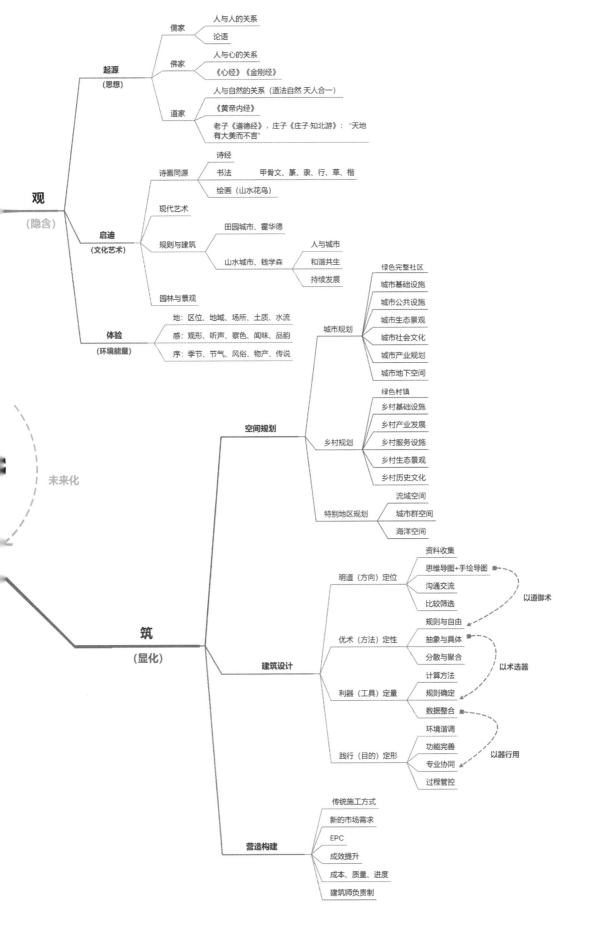

观
（隐含）

起源
（思想）

儒家
人与人的关系
论语

佛家
人与心的关系
《心经》《金刚经》

道家
人与自然的关系（道法自然 天人合一）
《黄帝内经》
老子《道德经》，庄子《庄子·知北游》："天地有大美而不言"

启迪
（文化艺术）

诗画同源
诗经
书法　甲骨文、篆、隶、行、草、楷
绘画（山水花鸟）

现代艺术

规则与建筑
田园城市、霍华德
山水城市、钱学森
人与城市
和谐共生
持续发展

园林与景观

体验
（环境能量）

地：区位、地域、场所、土质、水流
感：观形、听声、察色、闻味、品韵
序：季节、节气、风俗、物产、传说

未来化

筑
（显化）

空间规划

城市规划
绿色完整社区
城市基础设施
城市公共设施
城市生态景观
城市社会文化
城市产业规划
城市地下空间

乡村规划
绿色村镇
乡村基础设施
乡村产业发展
乡村服务设施
乡村生态景观
乡村历史文化

特别地区规划
流域空间
城市群空间
海洋空间

建筑设计

明道（方向）定位
资料收集
思维导图+手绘导图
沟通交流
比较筛选
以道御术

优术（方法）定性
规则与自由
抽象与具体
分散与聚合
以术选器

利器（工具）定量
计算方法
规则确定
数据整合
以器行用

践行（目的）定形
环境谐调
功能完善
专业协同
过程管控

营造构建
传统施工方式
新的市场需求
EPC
成效提升
成本、质量、进度
建筑师负责制

出版说明

Schematic sketch

The architect interprets and analyzes the needs of the project through the diagram, endeavors to find the key points and difficulties of the project, controls the design clues in the site, and surveys, walks, watches, experiences and feels the site information every time. The mind map plays several main roles in the process of the architectural design: deduction, induction, and hypothesis. Deduction is to start from finding historical materials, learning existing project materials from history, taking this as a basis, finding out their own design application methods and conceive design ideas. Induction is to understand and interpret the project, and find out the internal law through the observation and research of objective things. architecture and it's site are inseparable. The building does not exist alone, but is connected with the surrounding regional environment and has a special relationship. Therefore, the site of each project has its inherent characteristics, and the site generally has geographical and historical context. It is assumed that graphic thinking is a process of deliberation and selection of design schemes, which is convenient for creators to evaluate design schemes. At the beginning, the mind-map and freehand sketch were used for overall positioning of the project, then to local quantification, and finally for the details of the project to complete the work of finalization.

Watch . Build

The "complete building"includes both internal and external levels. In detail, architectural design is a process of"Inner View"and"External Construction":the "inner view"is partial to the concept and the inner, which separates perception from daily life. Namely,it is to observe the internal understanding and judgement for the site, to give up the simple attachment to the appearance, and to reach the state of superego, or the original form beyond nature."The inner view"is the process of research exploration and discovery of internal core elements, and excavating different spatial forms and demand patterns with an insight into the future."External construction"is the manifestation of concepts and the expression of meanings formed by using technical means. The inside and the outside are two inseparable sides: the inner view points directly to the core, and the outer construction externalizes semantics into the form. If we want to achieve a"complete building", return to the truth between the existence and reality of the building, and explore the original intention of the site.

图示草图

建筑师通过图示对项目需求进行解读和分析，发现项目的重点、难点，把控场地中的设计线索，每一次场地踏勘、游走、观看、体会，都是感受场地信息的过程。思维导图在建筑设计过程中主要起到几个作用：演绎、归纳和假说。演绎就是从查找历史资料开始，从历史中学习关于既有项目的资料，以此作为依据，找出适合自己的设计应用方法，构思设计思路。归纳就是认识和解读项目，通过对客观事物的观察与研究，找出内在规律。建筑与现实场地是密不可分的，它与周边地域环境相连并有着特殊的关系。每个项目地段都有其固有的特性。场地一般都具有独特地理性、历史性文脉。设计师会通过当地文化和自然对场地进行结合，找到场地中的隐含线索，发现自然中的规律，表达对文化与自然的一种尊重。假说是从灵感到构思的过程。建筑师会借助手绘草图把自己的构思给快速表达出来，将建筑设计方案进行抽象分析。假设的图示思维是一个对设计方案进行推敲与选择的过程，方便创作者对设计方案进行评估。思维导图与手绘草图刚开始是辅助设计师明确项目定位的方式，然后再到局部定量分析，最后落实在项目细节，完成定型工作。

观 . 筑

"完整建筑"包含了内与外两个层面。详细来说，建筑设计是一个"内观"加"外筑"的过程："内观"偏概念和内在，让感知从日常中脱离，即在自然本我状态下，观察审视内心对场域的理解与判断，放弃对表象的简单执着，而达到超我的状态，或者说超越自然的原始形态。"内观"就是对内在核心元素的研究性探索与发现的过程，并且在洞见未来中挖掘出不一样的空间形态与需求模式。"外筑"就是概念的显化与运用技术的手段形成意义的表达。它指向建筑的可见、可知、可用，形成再造的人工环境与自然的交互对话。内与外，是不可分割的一体两面："内观"直指核心，"外筑"将语义外化于形。如果要实现"完整建筑"，需要内外融合，即在建筑的有无之间、虚实之间回归本真，探寻场地的初心。

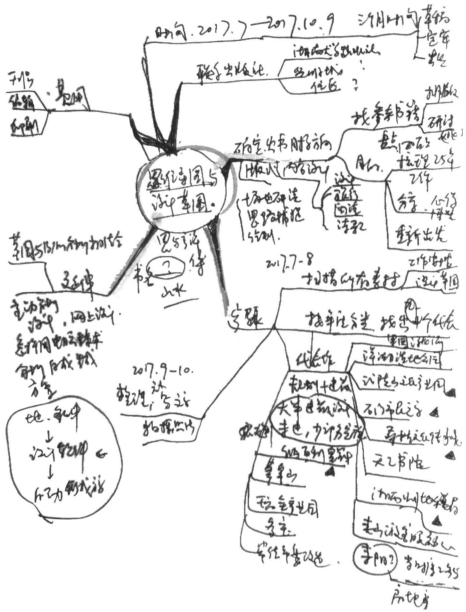

In addition,"the complete building"includes the design process of the whole process, the whole cycle, and the whole industrial chain, which is, through the unification of time and space, the cooperation between architecture and other supporting professions, and the coordination between architecture and users, to form a physical and spiritual building in harmony with Nature. Architecture becomes the continuation of Nature and life. I hope it is full of emotion and warmth. Architecture is an art of time, and design is also a practice of life.

此外，"完整建筑"包括全过程、全周期、全产业链的设计过程，即通过时间与空间的统一，建筑与其他配套专业的协作、建筑与使用者的协调，从而形成一个形神兼备、与自然和谐统一的建筑。建筑成为自然与生命的延续，希望它充满情感与温度。建筑是一门时间的艺术，设计也是对生命的一场修行。

Time and Architecture

In the concept of Philosophy in traditional Chinese culture, the heaven and earth are living bodies, and man is a small universe. Abstract concepts in cosmology need to be embodied in concrete material forms. Therefore, buildings, which are closely related to people's production and life naturally become the abstract incarnation of cosmic forms. When building houses, ancient people followed the example of the heaven and earth, trying to make the structure of the house consistent with the universe, pursuing the same origin and structure with the heaven , and in harmonious unity with nature. The concept of harmony in Chinese philosophical thoughts is reflected in conforming to the heaven and responding to humans , harmony and difference, and dual use. Through "figurative simulation" such as the direction, orientation, and decoration of the building, the cosmic schema of the unity of time and space and the philosophical thought of the unity of the heaven and man are reflected, to achieve a state of perfection. The traditional Chinese architectural space theory and Hetu Luoshu's Yin and Yang, five elements of the Shushu space structural forms are in total agreement. According to "Changes of Zhou: Cologne" ,"The river produces the map, and the Luo produces the book."

Architecture and Calligraphy

Unlike other pictorial characters, Chinese calligraphy is not only at the stage of representing symbols, but also has become a tool that can be used to express national grace. Chinese calligraphy and architecture have many similarities in line, plane-layout and structure, especially in the expression of abstract lines and spatial structures. The character itself is an extremely exquisite pattern. Chinese calligraphy art focuses on the penetration of the stroke structure and space, and the traditional architectural form also strives to create a spatial dialogue between the virtual and the real. Calligraphy is composed of basic elements of dots and lines, while architecture consists of walls, beams, columns, and other basic elements, both of which are constructing space.

Materials and Construction

Chinese traditional architecture extends along the extension of culture. In the cosmology of "Harmony between man and nature", architecture is a building of life centering on the nature. According to the Five-Element Theory, wood is symbolized by the

时空与建筑

在中国传统文化的哲学观念里，天地是一个活体，人是一个小宇宙。宇宙观中的抽象观念需要通过具体物质形式来体现。因此，和人们生产生活息息相关的建筑自然成为抽象宇宙形态的具象化身。古人在建筑房屋时，效法天地，试图让房屋与宇宙天地构造一致，追求与天地同源同构，与自然和谐统一。顺天应人、和而不同、执两用中，均体现了中国哲学思想中的和谐观。通过建筑的方位、朝向或装饰等"具象模拟"来反映时空合一的宇宙图式和天人合一的宇宙哲学思想，辨方正位，形意相生，使建筑与时空思想、艺术形式水乳交融，达到共臻其美的境地。中国传统建筑空间理论与河图洛书的阴阳五行术数空间结构形态是一脉相承的。《周易·系辞上》曰："河出图，洛出书。"

建筑与书法

中国书法不同于其他图示文字，历经千年沧桑已经成为一种可以用来表达民族美感的工具。中国书法与建筑在线条、平面布局以及结构上都具有很多的相通之处，特别是在抽象线条和空间结构表达方面有相似性。文字本身就是一幅极其精美的图案。中国书法艺术重在表现笔画结构与空间的渗透，而传统建筑形式同样力图营造虚与实的空间对话。书法以点线为基本构成元素组合，而建筑则以墙、梁、柱等基本构成元素组合，两者都在进行空间的构建。

材料与构筑

中国传统建筑顺着文化的延伸而延伸。在"天人合一"的宇宙观里，建筑是以自然为中心的生命建筑。在五行学说中，木以青龙为标志，方位为东，象征生气，是代表着生命的材料，中国建筑是一部"木头的史书"。因此，中国以木材为建筑材料，发展出了榫卯和斗栱等为连接手段的类框架结构，更容易获得大空间和灵活的布局，这恰好与现代建筑中混凝土框架结构不谋而合。从此，更多的混凝土以源自木结构的框架形式出现，逐渐成了主流的材料。

dragon, with its orientation in east, symbolizing vigor, and it is a material representing life. Chinese architecture is a"historical record of wood". Therefore, China has developed a quasi-frame structure with tenon and mortise and bucket arch as the connecting methods using wood as the building material, which makes it easier to obtain large space and flexible layouts, just coinciding with the concrete frame structure in modern buildings. Since then, more concrete has appeared in the structural forms derived from wooden structures and has gradually become the main-stream material.

The Metaverse and Creation

1992, the American novelist Neal Stephenson mentioned the word "Metaverse"in his science fiction named *Snow Crash*, the prefix "meta-" means transcendence, and the root- "verse" evolves from "universe", referring to the universe and the world. The metaverse is often used to describe the iterative concept of the future Internet, consisting of a persistent, shared, three-dimensional virtual space, a perceptible virtual universe. When a new wave of changes comes, the future and destiny of all people depend on the cognition of new things, and it is more necessary to work hard to understand the essence of things than to act in a hurry. Everyone needs to spend a lot of time and energy studying and thinking in order to truly understand the metaverse, especially to break through the barriers of thinking, master the"metaverse new thinking which is equal to technical thinking plus financial thinking and community thinking plus industrial thinking", so as to calmly cope with a series of new technological challenges in the future.

Multi-dimensional Extension, Wisdom, and Symbiosis

The architecture industry is facing unprecedented changes. The intellectualization of buildings has changed the constraints of previous production relations and stimulated new productivity through technology. The premise of intelligence is data, and design is the process and result of a project that expresses architectural ideas as graphics and models from scratch. It is the source of digitization and intellectualization for the entire architectural industrial chain.

元宇宙与创作

1992 年，美国小说家尼尔·斯蒂芬森就在其科幻作品《雪崩》中，提及"Metaverse"一词，其前缀"meta"意为超越，词根"verse"则由"universe"演化而来，泛指宇宙、世界。元宇宙通常被用来描述未来互联网的迭代概念，由持久的、共享的、多维的虚拟空间组成，是一个可感知的虚拟宇宙。当全新的变革浪潮来临时，每个人的前途命运取决于对新事物的认知，努力洞明事物本质比匆忙地行动更有必要。每个人都需要用大量的时间和精力进行学习和思考，以真正理解元宇宙，特别是要打通思维层面的壁垒，掌握"元宇宙新思维 = 技术思维 × 金融思维 × 社群思维 × 产业思维"，才能从容应对未来一系列的新技术挑战。

多维延展　智启共生

建筑行业面临着前所未有的巨变，建筑的智能化改变了以往的生产关系，通过科技激发新的生产力。智能化的前提是数据，而设计是把一个项目从无到有地用建筑构思具象表达为图形和模型的过程和结果，是整个建筑产业链数字化智能化升级的源头。

覆土型博物馆建筑研究——里耶博物馆

In order to alleviate the contradiction between economic development and environmental protection, human beings have put forward the concept of sustainable development, and formed design concepts such as ecological buildings, green buildings, and energy-saving buildings. Among them, earth-sheltered buildings have increasingly attracted extensive attention from industry insiders due to their advantages of low energy consumption, land-saving, and being conducive to optimizing the landscape environment. As an ancient and special architectural form, museum architecture not only has the functions of collecting, sorting out, researching, preserving and displaying human cultural heritages, but also carries more social significance.

Liye Town, Longshan County is located in the northwest border of Hu'nan Province and the northern end of Xiangxi Autonomous Prefecture. It is located on the border of Hu'nan Province and Chongqiong, as a major port of Youshui River and a key spot on the ways connecting Hunan Province with Sichuan Province and Chongqing.Liye has a long history and rich of culture. There are Neolithic ruins, three ancient city-ruins in the Warring States Period, the Western Han Dynasty and the Eastern Han Dynasty, as well as thousands of ancient tombs in the Warring-States Period, the Qin Dynasty and the Han Dynasty within the area of less than three square kilometers. The construction of the Liye Ancient City Museum is an important public construction project. The museum integrates "collection, research, and education"through the display of the historical relics of the ancient city of Liye and Qin Jian, not only undertaking the protection of the ancient city but also bringing out the cultural characteristics of west Hu'nan Province, and comprehensively displaying and explaining the splendid historical and cultural traditions of Chinese civilization to the public.

The general plane-layout borrows the concept of" a general design", where the building, the indoor and outdoor environment, and the display content are comprehensively considered and fully integrated. The design is guided by the principles of ecology and locality, and has determined the guiding ideology of "full respect for the environment"and by refining the essence of traditional architectural elements to form an architectural image full of oriental humanities, peace, simplicity and restraint. Efforts are made to protect and inherit the traditional architectural culture, while respecting the harmony between the nature and the environment. The layout of the site follows the overall planning road network layout. The building-layout will follow the topographic trend in a regular and stable way to divergent situation, and the buildings will be scattered along the slope. The central part of the site is a museum, the east side is a 2-story restaurant for visitors and a tourist service center, and the second phase on the west side is the office-building of the Institute of Archaeology. In the south, there is a front square along Hualong Road, where visiting vehicles are parked. There are entrances and exits to the office-building of the research institute in the west, through which staff vehicles enter.

The bridge: Inside the museum, a stone bridge surrounds the inner courtyard and forms a loop, implying that the museum is a link between history and reality.

The corridor: The exhibition is reasonably organized along indoor annular corridors to form a smooth visiting route.

为缓解经济发展与环境保护之间的矛盾，人类提出了可持续发展概念，并形成生态建筑、绿色建筑、节能建筑等设计理念。其中覆土建筑以其低能耗、节地、有利于优化景观环境等优点日益受到业内人士的广泛关注。而博物馆建筑作为一种古老而特殊的建筑形式，不仅具有收集、整理、研究、保存、展示人类文化遗产的功能，还承载着更多的社会意义。

龙山县里耶镇位于湖南省西北部，湘西土家族与苗族自治州北端，是湘、渝交界之地，酉水河上的重要码头，湖南通川、渝的咽喉。里耶历史悠久，文化源远流长，在该镇不足3000m² 范围内分布有新石器时代遗址，战国、西汉、东汉3座古城遗址，数以千计的战国、秦、汉古墓。里耶古城博物馆将通过对里耶古城历史文物以及秦简的展示，寓"收藏、研究、教育"于一体，使之既承担起保护古城珍贵遗产的功能，又能充分体现湘西地区文化特性，并向公众全面展示与阐释中华文明灿烂的历史文化传统。

总平面布局引入"总体设计"概念，即建筑与室内外环境及展示内容综合考虑，充分融合。设计以生态性原则和乡土性原则为指导思想，确定了"充分尊重环境"的理念，并通过提炼传统建筑元素精华，形成充满东方人文、平和、简洁、质朴、内敛的建筑形象和指导思想。保护和继承传统建筑文脉，同时尊重自然与环境协调。在场地平面布置中，遵循总体规划路网布局，建筑布局各体量将沿着地形、地貌趋势以一种规则、稳定到发散的态势展开，建筑沿坡地错落。用地中部为博物馆，东侧为 2 层对外餐厅及旅游服务中心，西侧二期为考古研究所办公楼。南向沿花龙公路设前广场，外来车辆在此停放。西侧研究所办公楼设出入口，内部车辆由此进入。

桥：博物馆内部一座石桥围绕内院，行成环路，暗喻博物馆是一座联系历史和现实的纽带。

廊：展览流线以室内环状连廊合理组织，形成顺畅的参观流线。

Research on Earth-Covered Museum Architecture - Lier Museum

The courtyard: The central courtyard and the small courtyards interspersed in the building constitute the exterior space of the building. The flowing water in the courtyard and the light and shadow changes of the bamboo form a restrained and subtle poetic painting. Due to the discovery of ancient tombs by Maicha, the excavation site will be preserved in the central courtyard in combination with landscape design, and the excavation time and display content will be marked so that the indoor courtyard will become an exhibition venue, making tourists feel like wandering in time and space.

The building is arranged on the central platform, spread out along the main axis, relatively concentrated, with a simple shape, corresponding to the mountain, and partially embedded in the mountain, forming a unified body with the base. The northern side of the building conforms to the mountain pattern, and a series of mountain and scenery viewing platforms are designed. The design does not amlessly pursue the grandeur of the building volume, but ingeniously treats it as a building ensemble , so as to achieve a harmonious unity with the surrounding mountains in scale. The shape adopts the sloping roof form with a similar slope to the mountain, giving people a sense of reality that the building grows in the mountain. The use of materials such as concrete with rough surface and veneer stone reflects the local style of the building ensemble . The original intention of the original ecological design is fully expressed by means of partial soil-cover and roof-greening.

Another major feature of the architectural form serves to create a rich external space. Inner courtyards, green roofs, and corridors allow the interior and exterior spaces to borrow from and penetrate through each other, and the rich elevation changes increase the sense of hierarchy of the space. Such an open and flowing external space provides a variety of possibilities for the user behaviors, reflecting humanistic care.

There are many historical relics around the Liye Museum. By introducing the concept of the eco-museum and using the museum construction to integrate the surrounding eco-tourism, natural landscape, and human history, the unique local culture of Liye is recommended to the world as a whole. The integrated eco-museum strategy will be more attractive than building a museum to display Qin Jian alone.

院：中心庭院与建筑中穿插的小型庭院，构成了建筑室外空间。庭院中的流水以及竹子的光影变化形成一幅内敛、含蓄的诗意画。因麦茶发现古墓葬群，在中心庭院中将结合景观设计把发掘遗址作原状保留，并标注发掘时间和展示内容，使室内庭院成为展场，令游客走在其中有漫游时空的感觉。

主体建筑布置在中部台地上，沿主轴展开，相对集中，形体简洁，与山地相对应，并在局部嵌入山体，与基地形成一个统一整体。建筑的北面顺应山势，设计出一系列的观山和观景平台。设计并没有一味追求建筑体量的宏伟，而是独具匠心地将其处理为建筑群落，从而达到了与周围山体在尺度上的和谐统一。形体上采用与山体坡度相近的坡屋顶形式，给人以建筑生长于山体中的真实感。表面粗糙的混凝土、贴面石材等材质的运用体现了建筑群体的乡土风格。通过局部覆土和屋顶绿化的方式，使建筑充分表达了原生态的设计初衷。

建筑形态的另一大特点是营造出了丰富的外部空间。内庭院、绿化屋顶和连廊等手法的使用使内外空间互相因借、渗透，丰富的标高变化更增加了空间的层次感。这样一种开放与流动的外部空间为使用者的行为方式提供了多种可能，体现出人文关怀。

里耶博物馆周边历史遗存众多，通过引入生态博物馆的理念，利用博物馆建设将周边的生态旅游、自然景观与人文历史进行整合，整体向世人展示里耶独特地方文化。整体化的生态博物馆策略将比单独建设博物馆展出秦简更有吸引力。

Inherit

朱明未承夜兮。——《楚辞·招魂》。注："续也。"

《说文解字》：奉也。受也。从手从卩从。

十二个重叠压缩的

春夏秋冬

磨砺智慧的年轮

手，描绘热土

心，群山回响

微光无言，等待答案

静水流深，凝聚期许

横山远水

显，内心风景

苍茫云海

隐，聚散虚空

建筑承载幸福，灵魂温暖燃烧

肩负阳光

探寻未知的世界

传承与延续
Inheritance and continuation

湘西武陵山文化产业园
Xiangxi Wuling Mountain Cultural Industry Park

建筑不仅满足人们对物质的需求，更是人们对精神需求的重要载体，建筑设计将地域性文化的思考直接表达出来，即建筑是凝固的文化艺术。由于民族交流与融合，民族区域特色文化逐渐消失。随着大众对传统文化认知的提升，优秀文化遗产仍有机会得以珍存。文化产业园区是后工业时代将文化与经济规律融合创新的产物，是艺术交流、文化生产、文化消费与城市公共空间结合拓展的发展模式。与工业产业园等产品生产型产业集群相比，文化产业园具有生态性、地域性、展示性、多元化等发展特点。文化产业园区的建设，可以带来开发带动地方观光旅游，强化地域文化特色交流，营销独特的地方特色，提升吸引公共投资竞争能力等有多重效益。

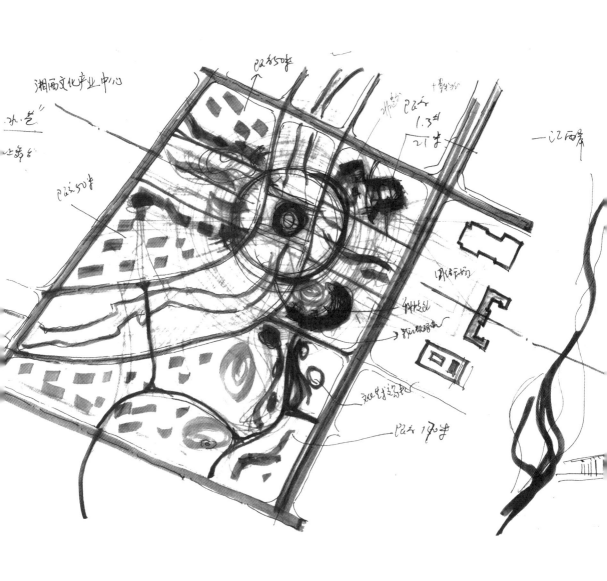

立春

冥冥甲子雨，已度立春时。
轻箑烦相向，纤絺恐自疑。

——唐·杜甫《雨》

文化产业园区地域特色创作实践
——湘西州武陵山文化产业园规划及建筑设计

文化遗产是不可再生的珍贵资源。由于民族交流的融合，民族区域特色文化逐渐消失。随着地域特色文化的觉醒，优秀文化遗产仍有机会得以珍存。然而，如果不能从产业集聚及旅游资源整合角度入手，只是单一地建设独立博物馆或艺术中心，文化产业园区往往面临着建设分散，运营成本高，无法发挥集群带动示范作用的问题。湘西土家族苗族自治州（以下简称湘西或湘西州）更是具有丰富的非物质文化遗产留存。"品类繁多、环境优渥、价值独特、传承濒危"成了湘西州非物质文化遗产保护的现状，因而各种类型的非物质文化遗产及物质文化遗产的保护展示，更是体现了湘西州文化产业园的重要内容。

1. 项目概况

湘西武陵山文化产业园位于湘西经济开发区核心区。武陵山文化产业园未来既是湘西自治州首府吉首举办 60 周年建州庆典场所，更是湘西州现代化水平和展示地域文化形象的窗口。湘西州政府行政服务综合楼前广场向西延伸形成城市轴线，文化产业园位于行政轴线上，地理位置优越。基地由已建成的吉凤路、丰达路两条开发区的主干道围合，湾溪河自西南向东北从基地中部穿过。基地以湾溪河为界，地势分别向东和向西逐渐升高，局部落差高达 20 m。

2. 规划理念

设计从湘西优美的山谷、碧水、廊桥等独特的景观中汲取灵感，通过现代设计手法，力求营造文化产业园"谷韵湘西、凤翔乾州"的城市景观大格局。

谷：湘西的山——奇在势，湘西的谷——绿在深。这次规划中，高层建筑与底层建筑形态组合，效仿湘西的山高谷低，进行高低错落布局，形成建筑块面丰富的自然姿态。设计力求从大场面的塑造中寻找湘西的自然风采，使人在其中穿行，犹如置身于湘西优美的山峦峡谷间，感受湘西大山的气势。

水：湘西是以山水秀美著称。规划基地有湾溪河自西南向东北从基地中部穿过，水资源丰富。规划保留并适当扩大湾溪河水面，并在两轴交界处形成以水为主题的节庆广场，还通过设置下沉式水景广场、音乐水景喷泉等现代高科技产品，以水为媒介，表现湘西的民风民俗。

桥：湘西的桥与百姓息息相关，同时也是湘西文化的重要构成部分。设计中桥不仅作为湾溪河两岸的通道，也是各类建筑形体组合与各类空间组合中不可或缺的要素。在桥的构造下，市民生活变得更加立体，更加多元。桥的概念在本次设计规划及建筑单体中得到了充分的诠释。

3. 方案设计

（1）Mangado 先生方案：苗寨、吊脚楼、巨石

博物馆、非遗馆设计：建筑的规划来源于传统苗族村落布局，一系列自由的单体由线性体量连接围合，创造出灵动的苗寨空间，在传承传统文化的同时也提供了近人的建筑尺度，创造出宜人的乡村景观。漂浮在公园里的单体象征着湘西州峡谷中分布的嶙峋巨石，棱角分明，层层叠叠。这些单体用半透明的玻璃围合，内部灯光亮起时，像发光的石头在诉说着自强不息的民族精神。它们也可以从外部用更加微妙的照明方式点亮。这种方式代表了当今的时代和人民向祖先创造的民族的传统和文化之根致敬。这些体块的表皮也可以作为投射关于民族遗产和习俗片段的大屏幕。

文化艺术中心设计：文化艺术中心便是这历史文化的延续，展示着当今湘西州的科技和艺术力量。广场上两组建筑是时间的两极，分别代表过去和未来，而广场是调和两极的中心。这座建筑延续了博物馆综合体的一些特点，用碎块化的形体构成建筑，同时也赋予这些碎片单体以岩石的象征。在这里，其外在形式上与博物馆相连续，内在形式空间采用不同的生成逻辑。建筑分为两个功能不同的区域：数码影视中心和文化艺术中心。这两个部分各有不同的主入口，内部是连接起来的。设计中两个部分都意在创造一种有机的自然空间，从而引发了拥有双层立面的形体。外层面向室外的较开放，由垂直的仿木金属板组成，可以实现遮阳；内层围合了内部空间。这组建筑与博物馆岩石的封闭性喻义不同，力求与所在的文化产业园实现更多的对话。可以这么说，这座建筑的价值在于室内外的紧密连接。

（2）设计院方案：湘西的山水意象、岩石梯田、背篓

博物馆、非遗馆设计：

"仁者之山"的场域意象，博物馆与非遗馆是以展示湘西州独特文化为主的博物馆，博物馆的"巨石"主体隐喻了稳定的、可信赖的山，通过层叠而上的覆土与延绵弯曲的非遗馆相连，代表着吉首人民和山一样平静，一样稳定，不为外物所动，同时通过各种展示表达了吉首人民像群山一样向万物展开双臂，宽容仁厚，不役于物，不伤于无，不忧不惧，始终矗立不变，包容万物。

复合的表皮肌理

建筑整体以山石为喻义，在采用了当地石材与现代玻璃为主要表皮材料之外，同时尝试将湘西独特的纹样图案融入之上，使建筑内的光外透，形成整体浑厚大气，又不乏湘西特有的吉祥浪漫的形态意向。

文化艺术中心设计：

"湾畔明珠"的形态构成设计以水的意向来反映建筑内涵，建筑造型如流水般流畅、圆润，体现出现代电子科技与多媒体的流动性及形态的不确定性。两个建筑主体近似椭圆形，像两个明珠镶嵌在湾溪河之畔。建筑基底采用梯田的形意，更增添建筑地域文化性。设计通过架空的处理方式，将单体建筑烘托出通透轻盈感，仿佛白云漂浮于恬静的水田之上，远远望去，如诗如画。

传统的表皮肌理

湘西州是一个少数民族聚集的地区，民族文化沉淀于湘西群山之中。设计将传统湘西文化书写在建筑之上，传承并发扬。数码影视馆将湘西苗族特有的银质花冠元素提炼融入外观设计中，夜幕降临，在湾溪河畔，熠熠生辉。

4. 结语

招标设计至施工图完成之前，设计团队通过多次的现场踏勘交流，与外国设计咨询团队密切合作过程中，也从不解逐渐到了解最终到欣赏的过程中对不同设计思路上有了深刻的体会。西班牙建筑大师对地域性建筑的创新性思考和对建筑内外部空间光线的完美表达，让人耳目一新。而国内设计方更多地受到建设工期及前期设计条件不足的影响，导致设计方案的反复修改。湘西自治州文化产业园项目作为湘西州的重点工程得到了业主及施工方的高度欣赏和支持，后期将绿色建筑设计及 BIM 引入了初步设计阶段也有效提升了项目的技术含量和完成度。该案例对于未来此类型文化产业园规划和设计原创性与地域文化的结合方面有一定的参考价值。

Creative Practice of Regional Characteristics in Cultural Industry Park
—— Planning and Architectural Design of Xiangxi Wuling Mountain Cultural Industry Park

Cultural heritage is a precious non-renewable resource. Due to the integration of ethnic exchanges, the characteristic culture of ethnic regions has gradually disappeared. With the awakening of regional characteristic culture, excellent cultural heritages still have the opportunity to be preserved. However, with a single independent museum or art center and not being able to think from the perspective of industrial agglomeration and the integration of tourist resources , cultural industry parks are still challenged by the problems of scattered construction, high operating costs, and disability to play the role of cluster demonstration. Xiangxi Prefecture has abundant intangible cultural heritages. "Various types, excellent environment, unique value and endangered inheritance" have become the current status of intangible cultural heritage protection in Xiangxi Autonomous Prefecture. Therefore, the protection and display of various types of intangible cultural heritages and tangible cultural heritages have become an important part of the Cultural Industrial Park of Xiangxi Prefecture.

1. The Project Overview

Xiangxi Wuling Mountain Cultural Industrial Park is located in the core area of Xiangxi Economic Development Zone. It will be the venue for the 60th anniversary of the founding of the state in Jishou, the capital of Xiangxi Autonomous Prefecture, as well as a window to the modernization of the prefecture and showcase the regional cultural image. At present, the square in front of the administrative service building of the Xiangxi Prefectural Government extends westward to form the city axis, and the cultural industry park is located on the administrative axis with a superior geographical position. The base is enclosed by two main roads of the development zone, namely, Ji Feng Road and Feng Da Road, which have been completed, and the Wanxi River passes through the middle of the base from southwest to northeast. The terrain gradually rises eastward and westward, with a local drop as high as 20 meters.

2. The Planning Concept

The design draws inspiration from the unique landscape of beautiful valleys, blue waters and bridges in west Hu'nan Province, and through modern design techniques, it aims to create an urban landscape pattern of"Valley rhymes with west Hu'nan Province and Fengxiang Qianzhou"in the Cultural Industry Park.

Valleys: Mountains of west Hu'nan Province feature the majesty, and valleys of west Hu'nan Province characterize the green color. In this planning, the combination of high-rise buildings and ground floor buildings form follows the high mountain and low valley in west Hu'nan Province, with a staggered layout of high and low positions, and a rich natural posture of building blocks. The design strives to reveal the natural style of west Hu'nan Province from the shape of the big scene, so that people can walk through it as if they were in the beautiful mountains and valleys of west Hu'nan Province and feel the momentum of lange mountains in west Hu'nan Province.

Rivers : Xiangxi is known for its beautiful landscape. The planning base has the Wanxi River running from southwest to northeast through the middle, offering abundant water resources. The planning retains and properly expands the Wanxi river, and build a water-themed festival square at the junction of the two axes. Through the setup of sunken water square, music water fountain and other modern high-tech products, with water as the medium, it showcases the folk customs of Xiangxi.

Bridges: Bridges in Xiangxi are closely related to people and are also important components of Xiangxi culture. Bridges not only serve as channels on both sides of the Wanxi River, but also as indispensable element sin the combination of various building forms and spaces. Upon the completion of bridges, the life of citizens has become more three-dimensional and diversified, and the concept of bridges is fully interpreted in this design planning and the single building.

3. Scheme design

(1) Mr. Mangado's scheme: Miao Villages, Stilted Buildings and Boulders

Museum, Intangible cultural heritage museum design: The planning of building comes from the traditional Miao village layout, A series of free monoliths are enclosed by linear volume connection to create a dynamic Miao village space, which inherits the traditional culture while providing a close architectural scale and creating a pleasant rural landscape. The monoliths floating in the park symbolize the jagged boulders scattered around the canyons of the Autonomous Prefecture. These monoliths are enclosed in translucent glass. When the internal lights are turned on, they resemble glowing stones that speak of the spirit of self-sufficient people. They can also be lit from the outside with more subtle illumination, a way that represents the times and people of today, paying homage to the traditions and cultural roots of the nation created by our ancestors. The surface of these blocks can also be used as a large screen for projecting video clips about the heritage and customs of the nation.

Culture and Art Center Design: The Culture and Art Center is a continuation of this history and culture, showcasing the technological and artistic power of Hu'nan Xiangxi Autonomous Prefecture today. The two groups of buildings on the square represent respectively the past and the future, and the square is the center that reconciles the two times. The building continues some of the characteristics of the museum complex, with fragmented forms composing the building. The fragmented monoliths are also given the symbol of rocks. Here the external form is continuous with the museum, while the internal formal space adopts a different generative logic, and the building is divided into two functionally distinct areas: the Digital Film and Video Center and the Cultural Arts Center. Each of these two sections has a different main entrance and is connected internally. Both parts of the design are intended to create an organic natural space, thus giving rise to a form with a double-layered facade; the outer layer facing the outside is open, consisting of vertical faux wood metal panels that allow for shade, and the inner layer endese the interior space. Unlike the closed metaphor of the museum rock, this group of buildings seeks to achieve more dialogues with the cultural industrial park in which it is located. In broef, the value of this building lies in the close connection between the interior and the exterior.

(2) The Scheme of the Design Institute : Landscape imagery, rocky terraces, bamboo basket in west Hu'nan Province
Design of museum and intangible cultural heritage museum.
The museum and the in tangible cultural heritage museum are museums that mainly display the unique culture of Xiangxi Prefecture. The main stone of the museum is a stauble and relialle wauntain, which is cannected with the rolling and bendly intangible Hewtage museum though the overlapping soil, representing that the people of Jishew are calm and stable as the mountain, and will not be mored by external thoungs. At the same time, through various displays, it expresses that the people of Jishou, like the mountains spreading their arms to everything, are tolerant and kind, do not serve things, and do not hurt anything. They are not worried or afraid, always standing unchanged and tolerate everything.
Composite skin texture.
In addition to using local stones and modern glass as the main surface materials, the building also tries to integrate the unique pattern of west Hu'nan Province into the surface , so that the light inside the building is transparent to the outside, forming an overall thick atmosphere with unique auspicious and romantic form of west Hu'nan intention.
Design of the Culture and Art Center.
The design of"the Pearl by the Bay"reflects the connotation of the building with the intention of water, and the shape of the building is as smooth and round as flowing water, reflecting the fluidity and uncertainty of modern electronic technology and multi-media. The two main bodies of the oval buildings resemble two pearls set on the bank of Wanxi River. The base of the building adopts the shape of terraces, adding to the regional culture of the building. The design of the building is light and transparent through the overhead treatment, as if clouds float on the quiet water-borne field, which is picturesque when viewed from afar.
Traditional surface texture.
Xiangxi is a region where ethnic minorities gather, and the culture of ethnic groups is steeped in the mountains of Xiangxi. The design of the building is written in the traditional Xiangxi culture, which is inherited and carried forward. The Digital Film and Video Museum incorporates the unique silver flower crown element of the Miao people in west Hu'nan Province into the exterior design, which glows at night on the banks of Wanxi River.

4. Conclusion
Before the whole bidding design to the completion of the construction plan, the design team obtained a profound understanding through a number of site reconnaissance exchanges, and a foreign design consulting team in the process of close cooperation, also in the process of gradually understanding the final appreciation of different design ideas. Spanish architectural masters' innovative thinking of regional architecture and the perfect expression of internal and external space light are refreshing. The domestic designers are influneced more by the construction period and the lack of early design conditions, resulting in repeated revisions of the design scheme. As a key project in Xiangxi Autonomous Prefecture, the project of Xiangxi Cultural Industrial Park has been highly appreciated and supported by owners and builders. The introduction of green building design and BIM into the preliminary design stage has also effectively improved the technical content and completion of the project. This case is of considerable reference value for the combination of originality in planning and design of this type of cultural industrial park and regional culture in the future.

湘西武陵山文化产业园（湘西博物馆＋文化艺术中心）

Wuling Mountain Cultural Industrial Park
(Xiangxi Museum + Culture and Art Center)

用地面积:
301400m²
建筑面积:
760460m²
容 积 率:
2.13
设计时间:
2013-2015
建造时间:
2015-2017

湘西武陵山文化产业园项目从地域文化特色与城市设计着手。从湘西优美的山谷、碧水、廊桥等独特的景观中汲取灵感,通过现代设计手法,力求营造文化产业园"谷韵湘西、凤翔吉凤"的城市景观大格局。

(下图:湘西武陵山文化产业园)

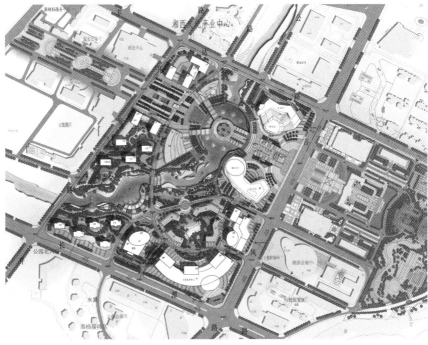

Site Area:
301,400m²
CFA:
760,460m²
FAR:
2.13
Design Time:
2013-2015
Construction Time:
2015-2017

Wuling Mountain Cultural Industrial Park of west Hunan features regional cultural characteristics and urban design. The design draws inspiration from the unique landscapes of beautiful valleys, limpid waters, and covered bridges in Xiangxi. With modern design techniques, it strives to create a large urban landscape pattern of the cultural industry park marked by "charming valley-Oriented Xiangxi and Auspicious Phoenix ".

(Below: Wuling Mountain Cultural Industrial Park)

方案 1：

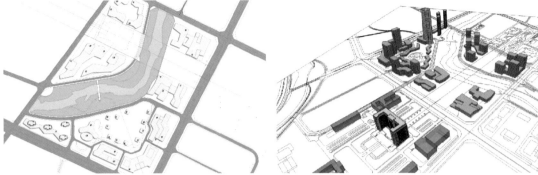

规划保留了地块原路网结构，只取消了四个场馆沿湾溪河部分道路。建筑布局比较规整，由于基地地块被湾溪河分割而缺少联系。方案缺少激动人心的空间大结构布局。

The plan retains the original road network structure of the plot, and only cancels a part of the roads along the Wanxi River for four venues. The building layout is relatively regular, and there is a lack of connection because the site is divided by the Wanxi River. The scheme lacks an exciting spatial and large structural layout.

方案 2：

规划调整了沿西北面沿湾溪河道路的走向，使之呈弧形与东南地块相连接，但建筑布局有些规整。

The planning adjusts the direction of the road along the Wanxi River along the northwest side so that it connects with the southeast plot in an arc shape, while the building layout is a little too regular.

方案 3：

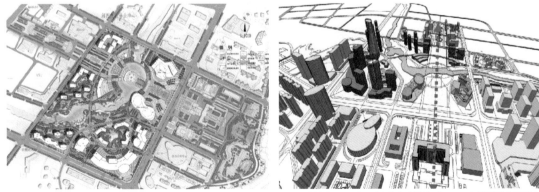

在方案 2 的基础上进一步调整和优化路网结构，调整建筑布局，使得建筑布局形态以州政府大楼为中心，形成翔凤来朝的城市大空间格局，最终形成本项目"一心两轴五片区"的整体规划构架。

On the basis of Scheme 2, the road network structure is further adjusted and optimized, and the building layout is adjusted so it is centered on the prefectural government building, forming a large urban spatial pattern of a Flying Phoenix , and finally forming the project's "one center, two axes, and five areas" overall planning framework.

最终方案（文化艺术中心）
Final scheme

本原　　　　　　冲积　　　　　　雕塑　　　　　　成形

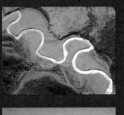

"湾畔明珠"

设计以水的意向来反映建筑内涵，建筑造型如流水般，流畅、圆润，体现出现代电子科技与多媒体的流动性及形态的不确定性。两个建筑主体近似椭圆形，像两个明珠镶嵌在湾溪河之畔。建筑基底采用梯田的形意，更增添建筑地域文化性。设计通过架空的处理方式，将单体建筑烘托出通透轻盈感，仿佛白云漂浮于恬静的水田之上。远远望去，如诗如画。

"Pearl by the Bay"

The design of"The Bright Pearl by the Bay"reflects the connotation of th building with the intention of water, and the shape of the building is a smooth and round as flowing water, reflecting the fluidity and uncertainty modern electronic technology and multi-media. The two main bodies of th oval buildings resemble two pearls set on the bank of Wanxi River. The bas of the building adopts the shape of terraces, adding to the regional culture the building. The design of the building is light and transparent through th

最终方案（博物馆）
Final scheme

本原

两馆分割

延绵起折

成形

"仁者之山"

建筑整体以山石为寓意，代表着吉首人民和山一样平静，一样稳定，不为外物所动，同时通过各种展示表达了吉首人民像群山一样向万物展开双臂，宽容仁厚，不役于物，不伤于无，不忧不惧，始终矗立不变，包容万物。

"Mountain of the Benevolent"

The whole building takes the mountain and stones as a metaphor, representing that the people of Jishou are as calm and stable as the mountain, and that they are not influenced by external things. They are tolerant and kind, do not serve things, and do not hurt anything. They are not worried or afraid, always standing unchanged, and tolerate everything.

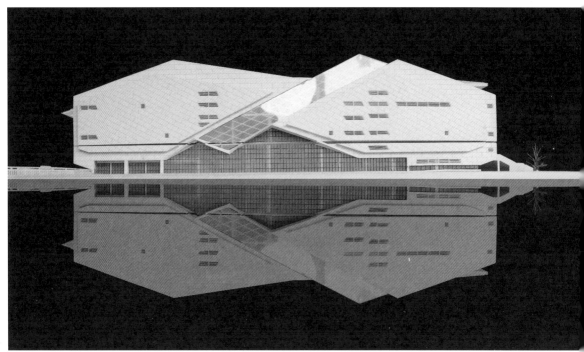

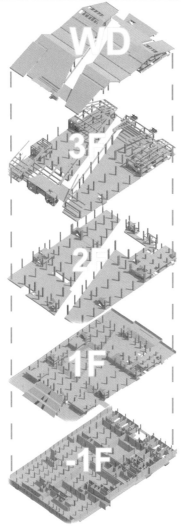

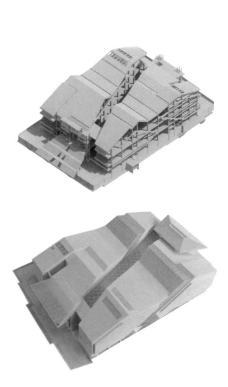

建筑体现城市特征，融入城市的
文化和历史，既是城市对外的展
示窗口，也是城市人的精神记
忆。设计中，城市文化的赋予是
建筑核心。

The architecture embodies
the characteristics of the city
and integrates the culture and
history of the city. It is not only
the window of the city, but also
the spiritual memory of the
people in the city. In design, the
embodiment of urban culture is
the core of the architecture.

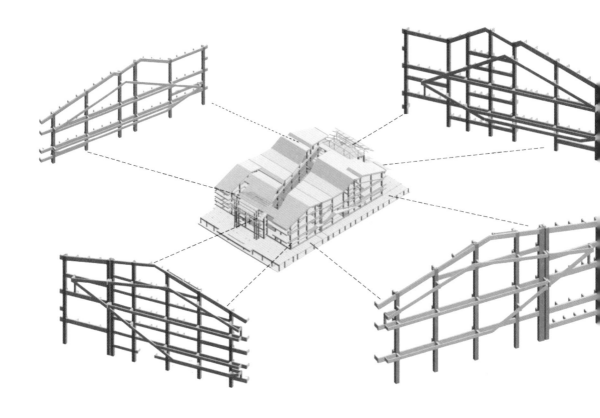

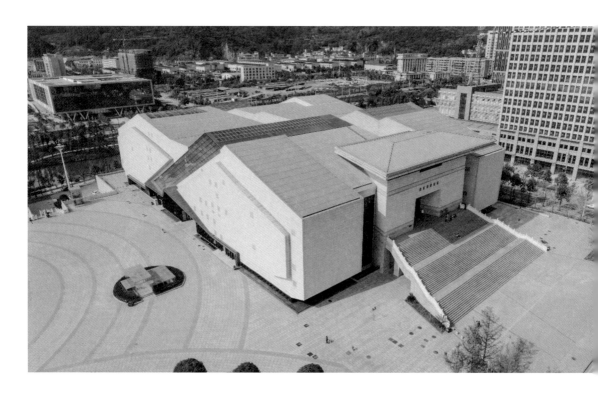

湘西独特的样纹图案"西兰卡普"，镂空的仿陶瓷挂板或其他金属板材附着在"延绵山体"的表皮上。当夜幕降临，建筑内的光柔和溢出，将建筑的"巨石"主体以山中之玉石为喻意。复合的表皮肌理融合了当地石材、现代玻璃。展现出整体的浑厚大气而又不乏湘西特有的浪漫之美。地上一、二、三层层高均为7.2m，地下一层层高为5.2m。地上首层为非遗馆，二、三层为博物馆。地下一层为车库及非遗馆的临时展厅。

Xiangxi's unique pattern marked by "Xilankapu", hollowed out imitation ceramic hanging boards or other metal plates are attached to the surface of the"extending mountain". The jade is a metaphor, and the composite surface mechanism integrates local stones, modern glass and transparency, showing an overall rich atmosphere and the unique romantic beauty of West Hu'nan Province. The height of the first, second, and third floors above ground is 7. 2m, and the height of the Basement 1 is 5. 2m. The first floor above the ground is the intangible cultural heritage museum and the second and third floors are the museum; the basement floor is the garage and the temporary exhibition hall of the intangible cultural heritage museum.

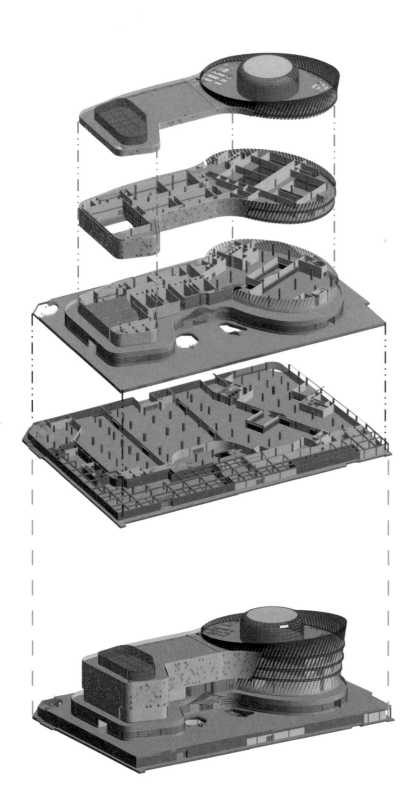

文化艺术中心的外表是通过数字化，将数字椭圆沿高程变化时水平方向发生旋转的过程作为项目的外观表现，并通过 BIM 技术的参数丰富立面形式，形成不同的开洞形式，方案表达丰富多样，有效保证了设计的准确性及设计精度，提高了设计效率。

玻璃幕墙进行开洞拆分。在建筑内部空间，传统二维设计很难想象内部空间组成，如梁高、梁宽等对实际净高的影响，玻璃幕墙是斜的，进一步增加了难度。外表皮结合数字化全专业 BIM 模型，利用 BIM 技术的可视化特征，在模型内直接对玻璃嵌板裁剪。

The outer facade is analyzed by digital methods, and the process of the digital ellipse rotating in the horizontal direction when the elevation changes is used as the external appearance of the project, The BIM technology is applied to enrich the facade form and to offer different opening forms, and the scheme expression is rich and diverse, effectively ensuring the accuracy and design precision and improving the design efficiency.

The glass panels are split through openings. In the interior space of the building, it is difficult for traditional two-dimensional designs to imagine the composition of the interior space, such as the influence of beam height and beam width on the actual net height, and the glass panels are inclined, which further increases the difficulty. It is scientific and reasonable to combine the digital full-professional BIM model and use the visualization features of BIM technology to directly cut the glass panels in the model.

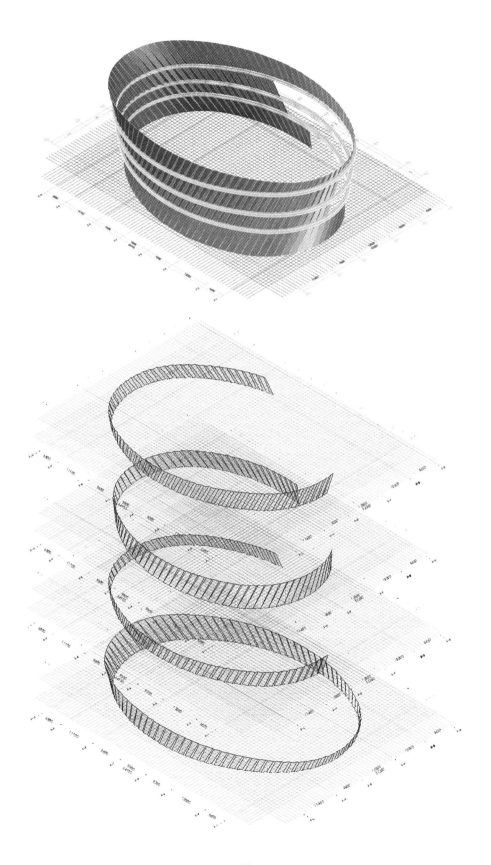

43

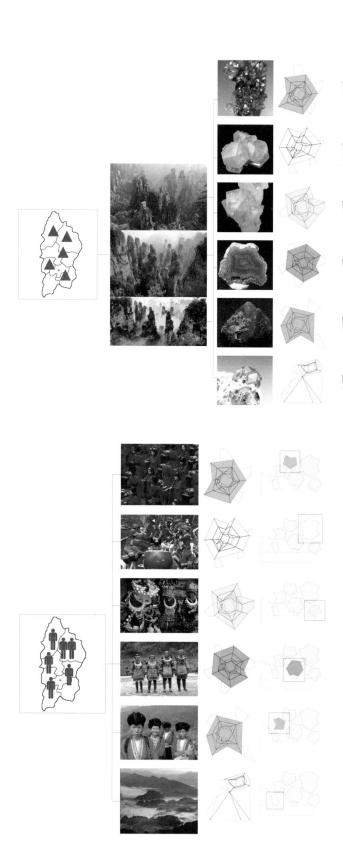

Brief Introduction to Mr. Mangdo's Scheme
Design strategy: endow the museum complex
with national characteristics, expressing national
unity, the interdependence of national cultures,
and the integration of water and milk. Endows
the museum complex with natural features, and
the architectural form evolves according to the
typical landform features of Hunan.

Mangdo 先生合作方案简介

设计策略：赋予博物馆综合体
以民族特质，表示民族团结，
民族文化相互依存，水乳交
融。赋予博物馆综合体以自然
特征，建筑形态根据湖南典型
地貌特征演变而来。

前期方案（博物馆）
Early stage plan

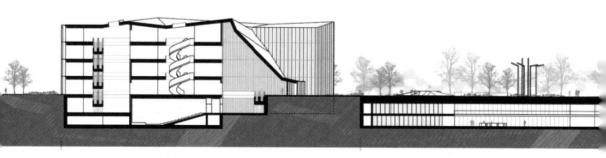

▲ 数码影视中心 / 文化艺术中心
Digital video Center / Cultural Art Center

▲ 文化广场东区 / 商店
Culture Square East / Sh

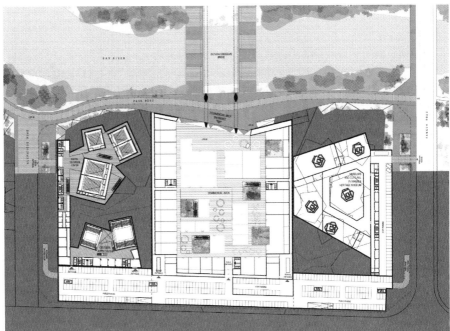

方案沿城市道路体量规划，与州政府呼应。内部体量采用了聚落式传统村落的变化转变。中轴线景观桥分为上下两层。创造性地运用了吊脚楼建筑形式，通过 2 层悬挑结构手段在建筑的底层架空出室外的活动场地。单体建筑微微抬高，整体效果仿佛是在公园里优雅作舞。人们可以在这片活动场地休憩游玩，也可以进行临时展览和非物质文化的制作体验。

This plan is schemed based on the planning of urban road volume, which agrees with the appeal of the prefectural government. The internal volume adopts the change and transformation of the settlement-style traditional villages. The central-axis landscape bridge is divided into upper and lower levels. It Creatively use the stilted building form through the two-story cantilever structure on the ground floor of the building to provide an outdoor venue for activities. The single body is slightly elevated, and the overall effect seems to be elegant as if they were dancing in the park. People can relax and play at this venue, and can also experience temporary exhibitions and the production of intangible culture heritage.

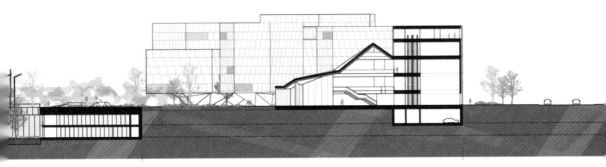

▲ 博物馆 / 非遗馆
Museum / Intangible Heritage Museum

涌现与转译
Emergence and translation

湖南常德石门市民之家
Hunan Changde Shimen Citizen's House

文化是城市的灵魂，"山水石门，诗意天成"的独特山水养育了一方人民，夹山、白云山是环绕石门县城的灵脉，那澧水就是穿越石门的经络，从城市文化中吸取养分。建筑艺术是有自身定位的，它具有地域性特征。地域特色如果在全球化趋势中能够很好地呈现出来，人们会更加珍惜。通过建筑的艺术性去保存城市的记忆，体现城市特色，提升城市的品质，展示城市的风貌，塑造城市的精神。市民之家建成以后，将为南城注入新生动力，成为综合型的历史文化休闲地，不仅服务市民，而且是旅游的目的地，成为石门整个城市的象征符号，百年后的历史建筑。如果说地域文化是城市的基因，那么地景式建筑也将成为基因传承的载体，水冲石起，时光留痕，文化在这里延续、转译。

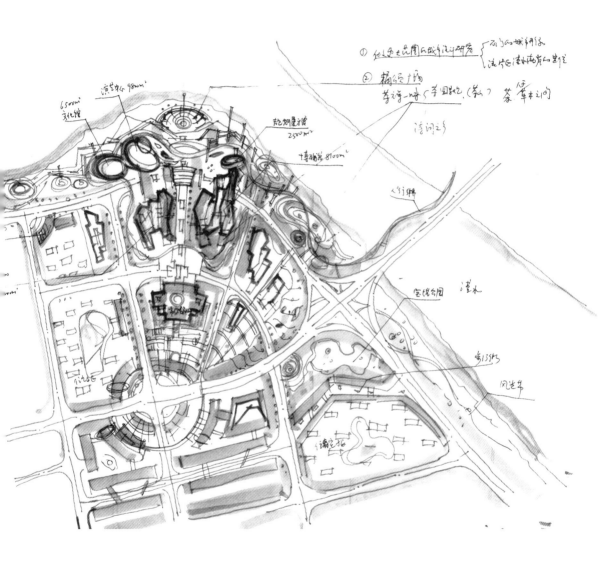

雨水

殆尽冬寒柳罩烟，熏风瑞气满山川。
天将化雨舒清景，萌动生机待绿田。

——宋·刘辰翁《七绝·雨水》

地域文化导向下的市民之家建筑设计
—— 常德石门县市民之家

伴随着中国"一带一路"的经济发展策略，中国社会也逐渐向国际化社会、知识化社会转变。在这样的时代背景下，我们不仅要关注社会生产力与经济结构的转变，还要更多的关注人文精神发展。在这样一个价值提升的新时期，我们不禁要对市民之家这类具有重要代表意义的功能复合型公共服务建筑提出关于 "市民之家怎样设计"与"市民之家如何彰显其城市意义"的提问。着重思考这两个问题能够让我们重新追根溯源地去探讨"市民之家"这类建筑的设计初心，并审视其发展过程中出现的问题。

1. 地域文化下的市民之家

各个地区因地形地貌不相同，长久的历史进程演化出不同的地域城市文化。市民之家作为城市必不可少的功能性建筑，设计时应在满足使用要求及功能性要求的同时兼顾当地的历史文化风俗，萃取城市的性格及人文特性，让建筑成为城市文化的载体，营造一个具有展示性的公共活动空间，弘扬和展示城市历史文化，同时营造建筑的独特性和美感，打造独具匠心的市民之家。宁乡县市民之家的建筑设计参考了由当地出土的国宝青铜器，建筑外立面上选用了石材幕墙作为装饰性材料，并在大片的石材墙面上刻印"四羊方尊"青铜器器身上的浮雕纹绘，以及政务服务入口上方的篆体"市民之家"彰显地域历史特色。

"天在巨星，飞驰湘北，轰然中开，而有石门。"石门是一个历史悠久，山清水秀，人杰地灵的地方。可以概括为三个特质：山水石门、生态石门、文化石门。"文化是城市的生命和灵魂，是城市的内核、实力和形象；城市是文化的凝结和积淀，是文化的容器、载体和舞台。"所以我们非常有必要，设计一个具备石门特有文化气质的市民之家。通过这样一个项目，去保存城市记忆，体现城市特色，提升城市品质，展示城市风貌，塑造城市精神，支持城市民展。市民之家建成以后，穿将为南城注入新生动力，成为综合型的历史文化休闲地，不仅服务市民，而且是旅游的目的地，成为石门城市的象征符号，百年后的历史建筑。同时，水是城市的血脉，我们想对包括市民之家项目在内的澧水两岸进行更大范围的城市设计，通过重塑一条河，达到更新一座城的目的。

2. 方案分析

我们这次设计不仅仅局限于"五馆两中心"的设计。开始从更大范围的城市设计及大数据分析、再到"五馆两中心"周边的城市设计、再到市民之家的规划设计及建筑单体设计、接着是景观设计，再到后面的展陈设计、场馆运营及品牌推广。先从城市的角度出发。项目位于南城临澧水位置，区域位置非常好，承东启西，可以说是城市的心脏。但是这个片区发展还不是很好，有几个原因：一是交通不便利，四桥的建立会起到很大的作用；二是电厂的影响，会带来一定的环境污染。只要解决好这些问题，市民之家这片区域是非常具备发展潜力的，甚至有可能成为新的城市副中心、文化高地。

我们提出了比较大胆的想法：既然电厂影响这么大，为什么不将它与市民之家片区隔离开呢？我们可以从南面的月亮湖引水，沿着夹山路形成一条水和绿地组成的隔离带，这样夹山路以北片区（我们不叫市民之家片区）就会独立开来，形成一个被水环绕的三角形绿洲，这样整个城市的格局会发生新的变化。市民之家片区的区位、环境、交通都是独一无二的，宜居宜商宜休闲，以文化引领的城市副中心就成为可能。

市民之家项目就是这个片区发展的起点，以它为中心，会形成几轴带几片区的城市空间格局（商业轴、景观轴、住宅区、教育区），市民之家是片区的凝聚点，这里聚集了人的活动，包括文化活动、运动休闲、慢行系统、游船观光码头、生态绿洲等等设施都围绕这个点沿着澧水两岸展开。

同时我们在想另外一个重要的问题，就是怎么去设计市民之家，什么样的市民之家才能代表石门？我们也参考了很多案例，比如永州市的"两中心"，将五馆放在一起形成一个大的建筑，但会体量很大，对公众缺乏亲近感；再比如益阳的"一园两中心"，十个建筑都是分散的，就没有中心，也不好管理；长沙的"三馆一厅"是做得比较好的，比较创新具有标志性，但绿化太少，我们的结论就是：

A. 空间尺度要开放亲民，不要做成一大块，建筑周边要设置更多的广场和休闲空间，更多的绿化空间，将公共建筑设计成公民建筑。

B. 相对要集约设计，有些功能能够组合的就组合在一起，动静分区，也便于管理。

C. 要有石门特色。也就是说什么样的市民之家才能代表石门？什么样的文化气质融入设计里面？是石门的橘子？还是石门的茶？经过讨论，大家认为石门的山水最能代表石门。壶瓶山一度是湖南最高峰，"湖南屋脊"，代表了"湖南高度"，澧水更是湘西北的母亲河之称。所以我们的设计采用的是"山水石门"的设计主题，通过"涌现"的设计理念。整个建筑群如水冲石起、时光留痕、诗意天成，如澧水河畔，绿色沙洲。文化在此演艺，柔软而流长。

3. 规划设计

在建筑形体和风格上，我们的建筑如山石一般起伏、延绵和地面景观相连，如在地生长而成。建筑立面采用如石头被水冲刷后的肌理，自然而具有韵律。由南往北，如水流向江河，刚开始是粗犷的、有棱角的石头体量和肌理，越往河边，形成的建筑形象是圆润的、流线型的。结合大的绿色生态环境，整个市民之家项目如一个综合性的文化湿地公园。

在建筑单体空间方面，我们也提出了很多创新，我们希望图书馆更开放一点，有意思的内部环境能吸引那些平时不爱看书的人也能参与进来；我们希望演艺剧院是半开放式的，舞台能打开，直接面对对面的山水环境，而不是一个完全闭塞的空间；我们希望公共广场是有活力的，微地形的处理能带来不一样的游园体验；我们希望滨河景观是接地气的，柑橘的种植采摘、茶禅一位的体验都都能融入进去等。

景观结合了我们的市民之家项目，以及宝塔公园地块。形成了一核、一园、五带、多节点的空间结构，动静分区，由心灵启示区、活力歌舞区、温馨剧场区、浪漫爱情区和修身养性区五个功能区组成，融入了茶禅文化、民歌文化、橘文化、石文化、兰文化等主题。临水空间我们设计了一个"橘子码头"，从长沙的"橘子洲头"到石门的"橘子码头"，更是一种品牌的打造。在景观设计里面，我们开展了很多专项设计，"海绵城市"理念的研究也施行，是这次设计一个重要课题。

4. 结语

石门县市民之家项目通过近 5 年时间的建设现已完工，工程模式从最开始的 3P 项目又改为 EPC 项目，中间经历了很多波折，但最终的建筑效果获得了较高的社会认可度，并因此带动了县城旅游经济的发展和相关产业升级。建筑通过对地域文化的汲取提炼，总结了开放、协同、共享的新型市民之家这一典型经验。未来将是县级文化综合体的新样本。

Architectural Design of Homes of Citizens under the Guidance of Regional Culture
—— Homes of Citizens in Shimen County, Changde

Along with China's "Belt and Road" economic development strategy, China's society has gradually shifted to an international, and knowledge-oriented society. In the context of such times, we should not only pay attention to the transformation of social productivity and economic structure but also pay more attention to the development of the humanistic spirit. In such a new period of value promotion, we can not help but put forward the issue about how to design the Homes of Citizens and how to highlight their urban significance for the the Homes of Citizens, which is a kind of functional compound public service architecture with important representative significance. Focusing on these two issues will enable us to trace back to the original design intention of such buildings and examine the problems during development process.

1. Homes of Citizens with Regional Culture

Due to different landforms in different regions, the long historical process has evolved different regional urban cultures. As an indispensable functional building in the city, the design of the homes of citizens should meet the requirements for use and functions, while taking into account the local historical and cultural customs, and incorporate the personality and humanistic characteristics of the city, so that the building will be the carrier of urban culture, hence creating a display of public space, carrying forward and display the history and culture of the city, and creating a unique architectural and aesthetic sense, to set up unique homes of citizens. The architectural design of the Homes of Citizens in Ningxiang County is based on the bronzes unearthed from the local treasures. The stone curtain wall was selected as the decorative material on the facade of the building, and the relief drawings on the bronzes of the"Four-Sheep Square Zun" were engraved on a large area of stone walls, as well as the seal character"Homes of Citizens" above the entrance to government service to highlight the regional historical characteristics.

"The sky is a giant star, flying in the north of Hu'nan Province, open in a sensation, and there is a stone gate." Shimen is a place with a long history, beautiful mountains and rivers, and outstanding people. It can be summarized as three characteristics: landscape Shimen, ecological Simen and cultural Simen. Culture is the life, soul, the core, strength, and image of a city. A city is the condensation and accumulation of culture, as well as the container, carrier and stage of culture. So, it's very necessary for us to design the homes of citizens with the unique cultural temperament of Shimen. Through such a project, it's aimed to save the memory of the city, reflect the characteristics of the city, improve the quality of the city, show the style of the city, shape the spirit of the city, and support the civil exhibition of the city. After the completion of the Homes of Citizens, the dome will add vitality to the southern city, and become a comprehensive historical and cultural leisure place, which not only serves the citizens, but also offers a tourist destination, a symbol of Shimen city, and a historical building years later. At the same time, water is the blood of the city. We want to design a wider range of cities on both sides of Lishui, including the Homes of Citizens project, by reshaping a river to achieve the purpose of urban renewal.

2. Scheme Analysis

Our design is not limited to the design of"five museums and two centers". We started from a wider range of urban designs and big data analysis, and then to the urban design around the"five museums and two centers", and then to the planning and design of the citizen-home and the single building design, followed by landscape design, and then to the following exhibition design, venue operation, and brand promotion. First of all, from the perspective of the city, the project is located in the southern city near Lishui, and the regional position is very good. It can be said that it is the heart of the city. But the development of this area is not very good, and there are several reasons: first, transportation is not convenient, the establishment of four bridges will play a great role to help; second, the impact of power plants will bring some environmental pollution. Only with these problems solved can the area of the Homes of Citizens be of great potential for development, and may even become a new city center and a cultural highland.

We have put forward a fairly bold idea. Since the power plant has such a great impact, why not isolate it from the Homes of Citizens area? We can divert water from the Moon Lake in the south and form an isolated area with water and green field along the Zhishan Road, so that the area north of Zhishan Road (we do not call it the Homes of Citizens area) will be independent, forming a triangular oasis surrounded by water, so that the entire urban pattern will change. The location, environment and transportation of the Homes of Citizens area are all unique, making it possible for people to live, do business, and relax in a culturally-led urban sub-center.

The Homes of Citizens project is the starting point of the development of this area. With it as the center, an urban spatial pattern (commercial axis, landscape axis, residential area, and education area) with several axes and several districts will be formed. The Homes of Citizens is the cohesion point of the area. People's activities gathered here, including cultural activities, sports and leisure, slow-moving systems, cruise ship sightseeing docks, ecological oasis and other facilities are all around this point, along both sides of the Lishui River.

At the same time, we are thinking about another important question, which is how to design a homes of citizens, and what kind of homes of citizens can represent Shimen? We have also referred to many cases, such as the "two centers" in Yongzhou City, which put five pavilions together to form a large building, but it will be very large and lack intimacy with the public; another example is the "one park and two centers" in Yiyang. ", ten buildings are scattered, there is no center, and it is not easy to manage; Changsha's "three halls and one hall" is relatively well done, more innovative and iconic, but there is too little greenery, Our conclusion is:

A. The spatial scale should be open to the people, not to be made into a large block. More squares and leisure space should be set up around the building, and more green spaces is needed so as to make public buildings civil buildings.

B. Intensive design is preferred. Some functions can be combined together, and dynamic and static partitions are also convenient for management.

C. The characteristics of Shimen must be brought about. That is, what kind of the homes of citizens can represent Shimen? What kind of cultural temperament shall be incorporated into the design?Shimen oranges? Or Shimen tea? It is believed that the landscape of Shimen can best represent Shimen. Huping Mountain was once the highest peak of Hu'nan Province, being famed as "The backbone of Hu'nan Province" and representing the "height of Hu'nan Province", Lishui is the mother river of northwest Hu'nan Province. Therefore, our design adopts the design theme of "landscape Shimen" , and the design concept of"emergence". The whole building ensemble , such as water rushing against stones, time leaving marks, and poetic genius, such as Lishui River, green sandbars, and culture here is soft and prolonging.

3. Planning and design

In terms of the building form and style, our buildings are like rocks, undulating and rolling, connected with the ground landscape, like growing on the ground. The building facade is colored like the texture of a stone after having been washed by water. It is natural and with rhythm. From the south to the north, it is like water flowing into the river, at first it is a rough and angular stone volume and texture, but the further to the river, the more rounded and streamlined the building image is formed. Combined with the large green ecological environment, the entire Homes of Citizens project is like a comprehensive cultural wetland park.

In the aspect of singular architectural space, we have put forward many new ideas. We hope that the library will be more open, and the interesting internal environment will attract people who usually do not like to read books as participants ; we hope that the theater of performing arts will be semi-open, and the stage can be opened directly facing the landscape environment instead of a completely closed space; we hope that the public square will be dynamic, and the treatment of micro-topography can bring a different sightseeing experience; we hope that the riverfront landscape is down-to-earth, and the citrus planting and picking, and the tea and Zen experience can be integrated into it.

The landscape combines our Homes of Citizens project and the Pagoda Park site. The spatial structure of one core, one garden, three belts and many nodes is formed, with dynamic and static partitioning, consisting of five functional areas, i.e. a spiritual revelation area, an energetic singing and dancing area, a theater area, a romantic love area, and a self-cultivation area, accommodating themes such as tea and Zen culture, folk-song culture, orange culture, stone culture, and orchid culture. In the waterfront space, we have designed an "Orange Pier" , from the "Orange Island" in Changsha to the "Orange Pier" in Shimen, which is a kind of brand building. In the landscape design, we have made many special designs, and the study of "sponge city" concept is also implemented, which is an important topic in this design.

4. Conclusion

The construction of the Homes of Citizens project in Shimen County has been completed in nearly five years, and the project mode has been changed from the initial 3P project to EPC project, which has experienced so many twists and turns. However the final architectural effect has gained high social recognition, thus having promoted the development of the tourist economy and related industrial upgrading in the county. The building summarizes the typical experience of an open, collaborative and shared new type of the Homes of Citizens by drawing and refining the regional culture. In the future, it will be a new sample of county-level cultural complex.

湖南常德石门市民之家

Hunan Changde Shimen Citizen's House

用地面积：
98351.4m²
建筑面积：
74762m²
容积率：
0.56
设计时间：
2015-2016
建造时间：
2017-2019

石门县市民之家位于常德市石门县东城区澧水北岸。总用地面积98351.4m²。是集展示、会演、政务、办公、商业、休闲、运动等功能于一体的文化公园。七个建筑单体按照使用性质分成相对独立的三组建筑，三组建筑既相对分散又有机组合，满足每个建筑的日照、通风、观景、人流疏散等要求。建筑地下一层，主要由地下停车场、食堂厨房、商业配套，以及设备用房组成。地上部分建筑层数为2~5层不等，建筑功能根据任务书要求灵活组织，充分利用公共空间共享和配套用房共用，达到使用的连续性和管理的高效性。

（下图：功能分区示意图）

Site Area:
98,351.4m²
GFA:
74,762m²
FAR:
0.56
Design Time:
2015-2016
Construction Time:
2017-2019

Shimen County people's home is located in Changde City Shimen County Dongcheng Li River north bank. The total land area is 98,351.4m². It is a cultural park integrating exhibition, performance, government affairs, office, business, leisure, sports and other functions. According to the nature of use, the seven monomer buildings are divided into three groups of relatively independent buildings. The three groups of buildings are both relatively dispersed and organically combined to meet the requirements of sunshine, ventilation, viewing and people flow evacuation of each building. The underground floor of the building is mainly composed of underground parking lot, canteen and kitchen, commercial supporting facilities, and equipment rooms. The number of floors of the above-ground part of the building varies from 2 to 5. The building functions are flexibly organized according to the requirements of the assignment book, and the sharing of public space and supporting rooms is fully utilized to achieve the continuity of use and the efficiency of management.

(Below: The schematic diagram of functional partition)

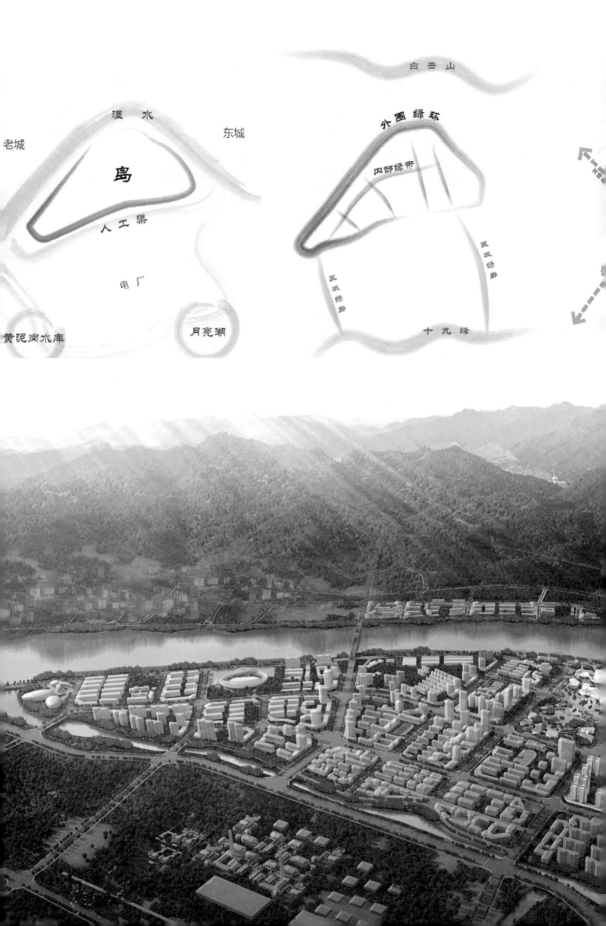

白云山

澧水　东城

老城

岛

人工渠

电厂

黄泥岗水库　　月亮湖

外围绿环

内部绿带

区域绿廊

区域绿廊

十九峰

规划采用"一核、二轴、五带、多片区"的规划布局结构：

"一核"是指：文化核心。

"二轴"是指：宝峰路城市发展轴线、梯云路城市发展轴线。

"五带"是指：澧水南岸滨水生态景观廊带、澧水北岸滨水生态景观廊带、夹山路滨水生态景观廊带、双宝路生态景观廊带、电厂路生态廊带。

"多片区"是指：文化中心区、商业商务休闲区、体育文化休闲区、各生态居住区等。

The planning adopts the planning layout structure of "one core, two axes, five belts, and multiple districts".

"One core" refers to the cultural core.

"Two axes" stand for the Baofeng Road Urban Development Axis and Tiyun Road Urban Development Axis.

"Five belts" refer to the waterfront ecological landscape corridor on the southern bank of Lishui River, the waterfront ecological landscape corridor on the northern bank of Lishui River, the waterfront ecological landscape corridor of Jiashan Road, the ecological landscape corridor of Shuangbao Road, and the ecological corridor of Power Plant Road.

"Multiple areas" refer to the cultural center area, the area of commercial business and leisure , the area of sports culture and leisure, various ecological residential areas, etc.

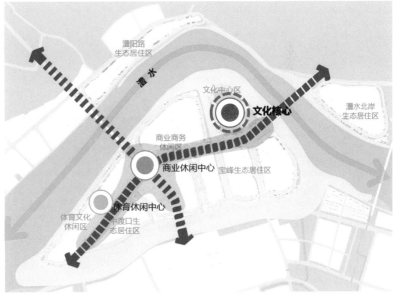

Colorful waterfront: Create diverse waterfront features and create vibrant waterfront space.

Bright core: The core leads the space radiation, shaping the core characteristics.

Sanshui Bend flow: Water as the medium, forming a variety of water space such as hydrophilic, Shuizhu, dry landscape.

Poetic corridor: Inheriting historical regional context and displaying diverse cultures.

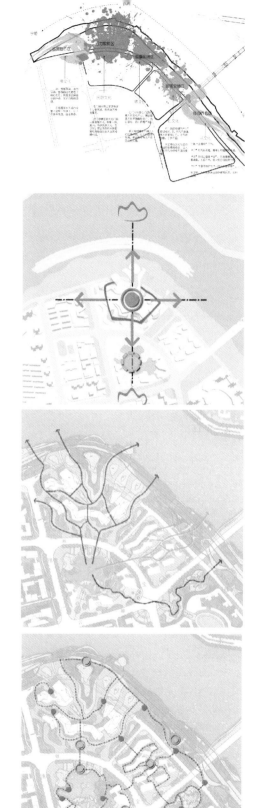

多彩水岸：塑造多样水岸风貌，创造活力滨水空间。

璀璨核心：核心引领空间放射，塑造核心特色风貌。

三水弯流：以水为媒，形成亲水、水渚、枯山水等多样水空间。

诗意走廊：传承历史地域文脉，展现多元文化。

"时光留痕"——设计如水冲石起、时光留痕、诗意天成，通过流水冲刷的自然形态，刻印到市民之家建筑群体中，建筑如生在水畔的滩石，或尖锐，或绵柔、建筑是澧水长河的印记，是石门历史的刻画，也是石门精神文化的展现。

"冲刷的建筑、流动的地景"——建筑不再仅仅是功能符号的展现，我们希望建筑是石门内在的精神图腾，是在地自然生长与承载时间印记的。我们的理解是建筑原初有一个基本形体存在，通过时间及文化激流的冲刷、打磨，最后成为所见的形体存在。这也是最本源的建筑形式，最接近合理的存在。

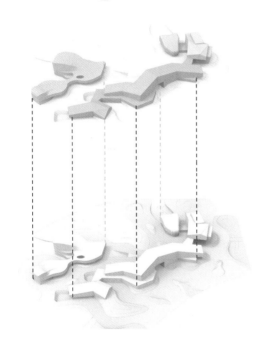

"Time imprint" — the design is like water flushing stone, time imprint, poetic Tiancheng, through the natural form of water flushing, engraved into the building group of citizens' homes, buildings are like beach stone born in the Lishui river, or sharp, or soft, the building is the mark of the Lishui river, is the portrayal of Shimen history, but also the manifestation of Shimen spirit and culture.

"Scour architecture and flowing landscape" — architecture is no longer just the display of functional symbols. We hope that architecture is the inner spiritual totem of Shimen, which grows naturally in the earth and bears the imprint of time. Our understanding is that there is a basic form of architecture in the beginning, and through the erosion and polishing of time and cultural torrent, it finally becomes the visible form. This is also the most original form of architecture, the closest to a reasonable existence.

"涌现"——基地的初始状态为一片空白荒地，我们通过多种要素分析，打造出一个核心点，而我们这个核心点是以水为主的景观广场，然后再通过水向四面八方蔓延，形成固定的建筑和景观形式。

"Emergence" — The initial state of the site is a blank wasteland. Through multiple factors analysis, we create a core point, and our core point is a landscape square dominated by water, and then spread in all directions through the water to form a fixed architectural and landscape form.

该项目利用 BIM 正向设计及大数据分析，对项目周边车流人流进行了定量模拟分析，并利用 BIM 模型进行了场地整体风环境、热环境模拟，计算和优化了建筑形态，提升整体环境的舒适度，打破传统的设计流程和协作模式，从而提高设计工作效率和水平，保障设计产品质量。通过推进 BIM 技术在设计施工中的应用，从而提高行业标准化、工业化、精细化管理的水平。BIM+VR，BIM+ 激光扫描等新技术的应用，更好地服务于项目，提高沟通效率。

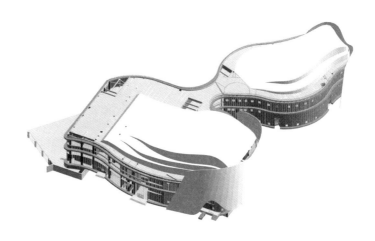

This project uses BIM forward design and big data analysis to carry out quantitative stimulate analysis on the traffic flow and people around the project, and use the BIM model to simulate the overall wind environment and thermal environment of the site, calculate and optimize the architectural form, and improve the degree of comfort of the overall environment. It will break the traditional design process and collaboration mode so as to improve the efficiency and level of design work and ensure the quality of design products. By promoting the application of BIM technology in redesign and construction, the level of industry standardization, industrialization, and refined management will be improved. The application of new technologies such as BIM+VR and BIM+laser scanning can better serve the project and improve communication efficiency.

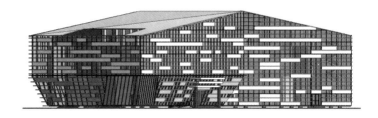

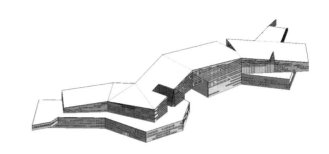

虚实与共享
Virtual reality and sharing

重庆腾讯双创社区
Chongqing Tencent Double Innovation Area

随着"大众创业，万众创新"热潮的进一步深入人心，"众创空间"一度跻身当下热政，成了承载创业梦想的"摇篮"和"基地"。相比地理空间的"实"，众创空间内的创业活动，同时也与空间之外的商业主体发生复杂的网络连接，这种商业网络是无限拓展的，具有无边界性和广域性，因此众创空间同时也是"虚"的社会网络空间。相比较传统的办公建筑，新型的办公综合体不再局限于某个地点，某个建筑形式，而是结合了休闲、娱乐、居住等的复合功能体。我们通过借助柯布西耶的设计理念，希望构建出灵活、自由，有创意和趣味的生态社区。这样一个具备双创气质的特殊项目，不仅满足了创新创业联合办公、创业解化、产业升级消费、国际化交流、新业态交互体验、商务社交、城市休闲等多元复合功能；还是"西南地区靓丽名片"、重庆"双创青年社区""一个时尚创意健康产业生态圈"以及"24 小时时尚活力新中心"。

雨催惊蛰候，风作勒花开。

——宋·陈棣《春日杂兴》

双创元素引领下的大型智慧社区研究

—— 腾讯双创社区（重庆高新）项目

当前我国经济社会正处于一个高速发展的关键时期，国内移动端和互联网的快速发展，促进了办公方式的深刻改革。相较于传统的创业园区，智慧社区不局限于某个确切的商业办公使用地点，也不固定于大型商业复合建筑设计使用形式，而是一个结合了商业办公、居住、休闲、娱乐等的新型现代复合商业办公使用功能体。其复合性、人性化、开放性、多元性等特征，说明了智慧社区已不是单一功能，而是相对复杂的社区形态。

国内对于智慧社区的建筑内外设计、空间功能设计、智慧服务系统和双创生态圈等研究相对较少。腾讯双创社区既是大型"智慧社区"的概念实践，又是"双创"中独具生命力的典型样板。因此，这里以腾讯双创社区为研究对象，通过大量文献的研究分析和国内现有实地项目的调研，总结出一套设计原理，为之后智慧社区在适应双创语境背景下的设计提供了一定的参考性。

1. 项目概况

腾讯双创社区（重庆高新）计划由区政府与腾讯众创空间联合筹建，融合产业办公、文创新商业为一体打造新型办公生活综合空间和新美学场景体验。以"智慧社区"为载体，将"双创"作为核心导向，通过构建创新创业生态圈来打造科技创新高峰，促进产城融合是发展高新技术产业和改造提升传统产业的重要基地和推进跨越式发展的新起点。

2. 设计理念

（1）策划

勒·柯布西耶曾言："住宅是用来居住的机器"，延伸至办公空间则为："办公室是用来工作的机器"。我们对重庆双创社区的设计内核，也由单纯的"联合办公社区设计"转变为"满足工作不断拓展的外延"，希望创造一个具备双创气质的特殊项目，构建出灵活自由、有创意和趣味的生态社区，能够让每个人在多元的办公氛围中穿梭。

（2）理念

在对项目的区域文化感地书写过程中，我们面临着一个关键性问题：具备什么样气质的双创社区，才能代表重庆？多次溯源与分析，我们认为认为是重庆的山、水、雾，以及重庆人乘风破浪，勇于创新的精神气质。结合具体环境分析，项目地处山城重庆，云雾环绕延绵起伏的歌乐山是大自然提供的视觉背景，辽阔宁静的彩云湖国家湿地公园轻涛抚岸。在这里，山、水与雾是有机的整体，山、水因雾而更名，雾因山、水而更灵。如何达到山、水、与云雾景观资源共享的同时不破坏大自然的和谐美是该项目规划所追求的目标。因此我们提出了"建筑追随自然"的设计理念，将建筑融于大自然。建筑不再是山、水、与云雾之间的屏障，相反，建筑是自然的一部分，并能够加强山、水与云雾的对话，从而达到与自然的和谐共生。

3. 规划设计

（1）手法

考虑到项目特殊的地形及地理位置，我们决定整个设计以一种特殊的方式，形成低层高密度的创业社区。设计从环境入手，通过综合分析用地周边环境因素，确立企业办公和商业的组成。整个建筑以一种独特的，符合在地建筑形态的院坝空间取得区别于一般商业效果的高端购物环境。将集中商业作为整个项目跳跃点，布置在地块的中心地下空间，特殊的形象点缀了整个线形空间。建筑物之间的间隙，层层退台，采用用铝合栅格围合，形成良好的视觉通廊，使景观向内廷渗透。整个地块的建筑物尽可能的贴近地面。设计有意将环境和绿化通过露台发展到三维的空间中去，使驻足于此的人们能够充分地享受到相互沟通、相互对话的不同高度、不同大小的空间。

（2）建筑

对于建筑的设计，我们则是结合"山地梯田"的层叠方式在建筑"内部"运用展示，同时植入绿色植被，达到城市、建筑、环境的有机结合，将室外园林景观引入到室内，以自然的植被，软化建筑的直线条，主富室内环境景观，改善室内气候条件，形成人性的环境氛围。在各层，为提升办公环境的质量，在建筑的中心内庭，根据房间平面布局，层层退台，形成共享边庭花园，为建筑增加了生气，亦改善了建筑的内部环境。

建筑立面采用渐变整列的方式，自下而上，大量装配式构建的大量重复应用，使建筑拥有沉稳的体量。中部随材质密肋的变细、玻璃颜色的变浅，视觉感逐渐轻盈。顶部构件更为纤细，使建筑向上的轻盈感更加强烈。平直的造型，给人向上、积极的形象；而渐变的立面不仅给人沉稳的体量感，也增强了建筑与人的亲和力。

（3）景观

开阔的视野和延绵的山景是城市赋予本项目得天独厚的资源，景观规划仍是以山、水为主题，使景观资源利用达到最大化。将清新大自然景观引入本地块，使人们充分体验自然的魅力。整个线性商业空间结合地块的形态搭建横向景观构架，设置以人文景观为特色的绿化、广场、亭台、步道，营造温馨和谐的社区氛围。纵向通过有机的组合建筑空隙将无限的景观融入内部，纵横相接，构成丰富多变的景观效果。

4. 智慧社区创新

（1）智慧装配

本项目采用装配式建筑，从方案设计初期就把装配理念贯穿项目全专业设计，项目单体建筑装配率为55%。主要通过综合装配率和工程造价的因素考虑，确定了竖向构件和主梁采用高精模板的方案。根据装配式设计方案，搭建项目整体模型，对结构模型中预制构件进行预制属性指定，开展整体结构计算、分析及优化，并根据计算结果对预制构件进行配筋。完成配筋后，通过BIM平台，对预制构件的钢筋、预制构件、机电模型进行冲突检测及避让调整。采用PKPM－PC进行预留孔洞设计，生产图纸绘制，构件清单统计，以指导后期生产及施工。

（2）BIM技术沟通智能

项目基于BIM技术对建筑进行施工仿真可视化处理，在前端有效的解决设计和施工中暴露的问题及隐患，避免现场工序冲突和作业返工造成的不必要浪费。同时，通过可视化模型，可以更加直观具体地对项目进行布局结构调整，实现智能高效、绿色环保的建筑设计目标。在以往建筑空间设计中，采光、遮阳、通风、排水等功能布局只能凭借设计人员的经验进行调整设计，缺乏客观科学的数据支撑。借助BIM技术构建三维立体模型，则可以实现构筑物环境的全方位模拟分析，确保光照、通风、排水等效果最优化，保证布局设计的科学性。在节能减排的建筑设计原则中，嵌入了更多智能化元素，以人为本，实现了人、自然、建筑的和谐统一，为人们创造出高品质、更宜居的办公生活模式。

5. 结语

在创客需求多元化、创业大环境日益复杂的背景下，智慧社区凭借智慧化、共享化、信息化、功能弹性化、全面化等特点，日益成为我国双创发展的强大驱动力，助推创新创业者成长。智慧社区既是新型办公空间的物理载体，同时也提供了创业的全流程服务，不是简单的地产模式，而是办公模式升级的产物，也是双创的另一种重要表现形式。未来在"双创"模式的大环境下，以共享型智慧社区为主的各种新型办公模式将得到广泛推动和大力发展。

Research on Large-Scale Intelligent Community Led by Innovation and Entrepreneurship Elements
——Tencent Innovation and Entrepreneurship Community (Chongqing High-Tech) Project

Economy and society in China are currently in a critical period of rapid development, and the rapid development of mobile and Internet in China has promoted a profound reform of office practices. Compared with the traditional business park, the smart community is not limited to a specific commercial office location, nor is it fixed to the architectural design of a large commercial complex . Instead, it is a new type of modern composite commercial office function that combines commercial offices, residences, leisures and entertainments. Its composite, humanization, openness, multiplicity and other characteristics indicate that the smart community is not a single function, but a relatively complex community form.

At present, there is relatively little research on the internal and external design of buildings, space function design, intelligent service system, and Innovation and Entrepreneurship eco-system of smart communities in China. Therefore, this paper takes Tencent's Innovation and Entrepreneurship community as the research object. Through a large amount of literature research and analysis as well as research on existing field projects in China, a set of design principles have been summarized to provide a basis for the design of Tencent's Innovation and Entrepreneurship community. This paper summarizes a set of design principles to provide some reference for the design of smart communities in the context of Innovation and Entrepreneurship in the future.

1. The project overview

The Tencent Innovation and Entrepreneurship Community (Chongqing) is planned to be jointly built by the regional government and Tencent Crowd Innovation Space, integrating industrial offices and cultural innovation business to create a new office life integrated space and new aesthetic scene experience. Taking "smart community" as the carrier, and "innovation and entrepreneurship" as the core orientation, it builds the ecological circle of innovation and entrepreneurship to create the peak of scientific and technological innovation and promote the integration of industry and city. It is an important base for the development of high-tech industries and the transformation and upgrade of traditional industries and a new starting point for promoting great-leap development.

2. Design Concept
(1) Planning

Le Corbusier once said: "Residence is a machine for living." Likewise, if we extend that to the office space: "Office is a machine for work ." The design core of Chongqing's Innovation and Entrepreneurship Community has also changed from a simple "joint office community design" to "meeting the expanding extension of work". We hope to create a special project with innovation and entrepreneurship features, build a flexible, free, creative, and interesting ecological community, and enable everyone to commute in a diversified office atmosphere.

(2) The concept

In the process of writing the perceptual texture of regional culture of the project, we are faced with a key problem: what kind of features of the Innovation and Entrepreneurship community can represent Chongqing? With origin-tracing and analysis of several times, we believe that they shall be Chongqing's mountains, water, fog, and Chongqing people who dare to ride the wind and waves in innovation. Combined with the specific environmental analysis, the project is located in the mountain city of Chongqing, surrounded by clouds and mists. The undulating Gele Mountain is the visual background provided by the nature. The vast and peaceful Caiyun Lake National Wetland Park is endowed with marvelous sceneries. Here, mountains, water and clouds are an organic whole. Mountains and water are renamed due to clouds and the clouds is more spiritual due to mountains and water. How to achieve the sharing of mountain, water and cloud landscape resources without damaging the harmonious grace of Nature is the goal of this project. Therefore, we have put forward the design concept of "architecture follows the nature." Integrating architecture with the nature, architecture is no longer a barrier among mountains, waters, and clouds. On the contrary, architecture is a part of the nature and can strengthen the dialogue among mountains, waters and clouds, so as to achieve harmonious coexistence with the nature.

3. Planning and design
(1) Approach

Considering the special topography and geographical location of the project, we decided that the whole design would be in a special way to form a low-rise high-density entrepreneurial community. The design starts from the environment and establishes the composition of corporate offices and commerce through comprehensive analysis of environmental factors around the site. The unique building is in line with the local architectural form of the dam space to achieve a high-end shopping environment, which is different from the general commercial effect. The centralized commercial space as the jumping point of the project is arranged in the central underground space of the site, and the special image embellishes the whole linear space. The gap among buildings is retreated layer by layer and enclosed by aluminum lattice, forming a good visual corridor and allowing the landscape to penetrate through the inner court. The buildings on the site are as close to the ground as possible. The design intends to develop the environment and greenery into a three-dimensional space through the terrace, so that people who stop here can fully enjoy the space of different heights and sizes for communicating among themselves.

(2) Architecture

For the design of the building, we combine the cascading method of "mountain terraces" to display the "interior" of the building, while implanting green vegetation to achieve the organic combination of the city, architecture and environment. Outdoor gardening is installed inside the building to soften the straight lines of the building with natural planting, enrich the indoor environmental landscape, improve the indoor climate conditions, and form a human environment. On each floor, in order to improve the quality of the office environment, the central inner court of the building is retired in layers according to the layout of the rooms to form a shared side garden, which adds vitality to the building and improves the internal environment.

The building facade adopts a gradual and integral approach from the bottom to the top, with a large number of assembled structures repeatedly applied to give the building a calm and steady character. The middle part of the building becomes lighter with the thinning of the dense ribs and the lighter color of the glass, while the top part of the building is more slender, giving the building a stronger sense of upward lightness. The flat and straight shape impresses people with an upward and positive image, while the gradual change of the facade not only gives people a sense of calmness but also enhances the affinity between the building and people.

(3) Landscape

The open view and ever-extending mountain sceneries are the unique resources given by the city to this project, and the landscape planning is still based on the theme of mountain and water to maximize the use of landscape resources. The fresh natural landscape is introduced into the site so that people can fully experience the charm of the nature. The whole linear commercial space is combined with the shape of the plot to build a horizontal landscape framework, setting up greenery, squares, pavilions and walkways featuring humanistic landscape to create a warm and harmonious community atmosphere. Infinite landscapes are integrated into the interior through the organic combination of building gaps, which connects vertically and horizontally to form a profound and varied landscape effect.

4. Intelligent Community Innovation

(1) Smart assembly

The project adopts the assembly-type building, and the assembly concept is carried through the whole professional design of the project from the early stage of scheme design. The assembly rate of the single building of the project is 55%. Based mainly on the comprehensive assembly rate and project cost, the scheme of adopting the formwork of high precision for vertical members and main beams were determined. According to the assembly design scheme, the overall model of the project was built, prefabricated properties were specified for the prefabricated members in the structural model, overall structural calculation, analysis and optimization were carried out, and reinforcement was allocated to the prefabricated members according to the calculation results. After completing the reinforcement allocation, through the BIM platform, conflict detection and avoidance adjustment were carried out for the reinforcement, precast elements and electromechanical model of precast elements. PKPM - PC was used to design the reserved holes, draft the production drawings and make the list of components guide the later production and construction.

(2) Communication intelligence of BIM technology

The project is based on BIM technology for construction simulation visualization of the building, which can effectively solve problems and hidden danger exposed in the design and construction process at the front end, and avoid unnecessary waste caused by process conflict and operation rework on site. At the same time, through the visualization model, the layout structure of the project can be adjusted more intuitively and concretely to achieve the goal of intelligent, efficient, green, and environmentally friendly architectural design. In the previous architectural space design, the functional layout of lighting, shading, ventilation, drainage, etc. can only be adjusted by the experience of designers, lacking the support of objective scientific data . With the help of BIM technology to provide a three-dimensional model, it is possible to achieve a comprehensive simulation analysis of the building environment to ensure the optimization of light, ventilation, drainage, and other effects, and to ensure the scientific nature of the layout design. In the architectural design principle of energy-saving and emission-reducing, more intelligent elements are embedded. The harmony and unity of people, nature, and architecture are realized so as to create a high-quality and more livable office life mode for people.

5. Conclusion

Under the background of diversified needs of makers and an increasingly complex entrepreneurial environment, smart communities, with the intelligence, sharing, informatization, functional flexibility, and comprehensiveness, are increasingly becoming a powerful driving force for the development of Innovation and Entrepreneurship in China and boosting the growth of innovative entrepreneurs. It is not a simple real estate model, but a product of upgrading the office model, and another important expression of Innovation and Entrepreneurship. In the future, under the environment of "Innovation and Entrepreneurship", various new office models, mainly shared intelligent communities, will be extensively promoted and vigorously developed.

腾讯双创社区（重庆）高新项目

Chongqing Tencent Double Innovation Area

用地面积:
25960m²
建筑面积:
124363m²
容积率:
3.0
设计时间:
2017-2018
建造时间:
2019-2020

腾讯双创是社区采用 EPC 总承包模式，用地面积约 25961m²，容积率小于等于 3.0，地上 6~8 层，地下 3 层。拟建地上总建筑面积 77883m²（商业 39588m²，办公 38394m²），地下总建筑面积：商业 7632m²，车库 38847m²。 总建筑面积 124363m²，项目总投资约 11.3 亿元。

（下图：功能分区示意图）

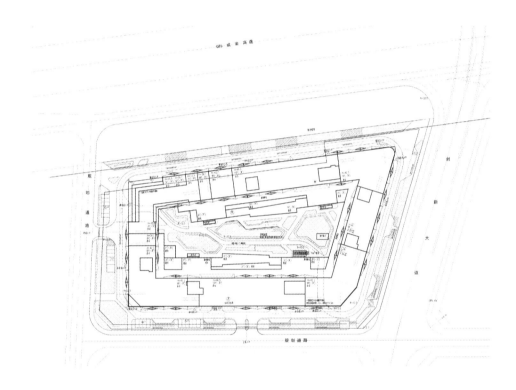

Site Area:
25,960m²
GFA:
124,363m²
FAR:
3.0
Design Time:
2017-2018
Construction Time:
2019-2020

Tencent Innovation and Entrepreneurship Town adopts engineering procurement constitution(EPC).Coverage: 25,961m², Floor area ratio: ≤ 3.0, Levels: 6-8, Basement levels: 3, Gross floor area above ground (proposed): 77,883m²(business: 39,588m²; office: 38,394m²). Gross floor area under ground (proposed): 7,632m² for business, 38,847m² for garage. Construction floor area: 124,363m². Total investment: 1.13 billion yuan (about $ 162 million)

(Below: The schematic diagram of functional partition)

具备什么样气质的双创社区，才能代表重庆？我们认为认为是重庆的山、水、雾，以及重庆人乘风破浪，勇于创新的精神气质。

青山、秀水、与云雾景观资源共享的同时，不破坏大自然的和谐美是该项目规划所追求的目标。因此我们提出了"建筑追随自然""山、水、青、秀"的设计理念。将建筑融于大自然，建筑不再是山、水、与云雾之间的屏障，相反，建筑是自然的一部分，并能够加强山、水与云雾的对话，从而达到与自然的和谐共生。

What kind of features of Innovation and Entrepreneurship community can represent Chongqing? We believe that it is Chongqing's mountains, water, fog and the spirit of Chongqing people who ride the wind and waves with courage in making innovation.

While green mountains, limpid waters and cloud landscape resources are shared, it is the goal pursued by this project without destroying the harmonious grace of the nature. Therefore, we put forward the design concept of "building follows nature" and "Mountain, Water, Greenery and Grace". When incorporating architecture into the nature, architecture is no longer a barrier among mountains, water and clouds. On the contrary, architecture is a part of the nature and can strengthen the dialogue among mountains, waters and clouds to achieve harmony with the nature.

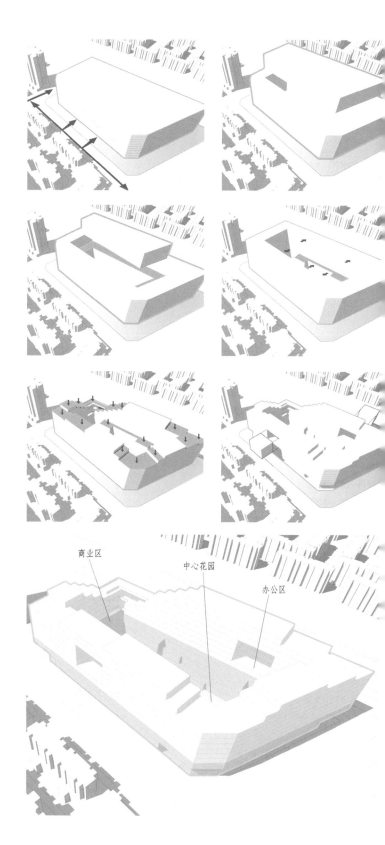

商业区　　中心花园　　办公区

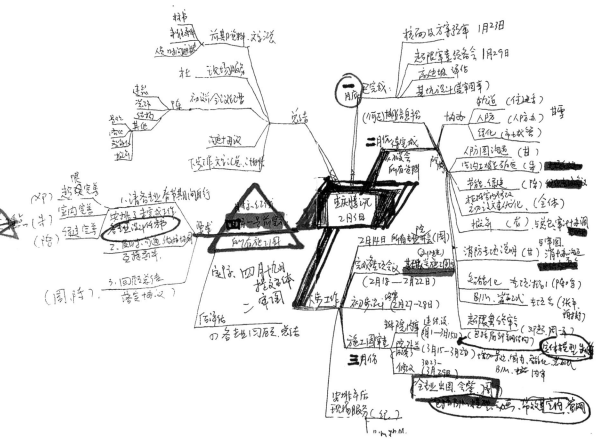

方案从 5 个方面最大限度地体现了创意文化社区的聚集发展特性。利用率大；场地内部空间丰富，可较好地吸引人流进入其中；商业空间与办公空间呈竖向分布，商业集中在 1~2 层，办公空间集中在 4~8 层，避免了相互之间的干扰，但又相互连接，使用便利。引入中庭，内部空间丰富，增加了景观界面。整体呈围合之势，与外部空间环境形成复合界面，并限定内部空间范围，实现与城市环境相融。

The project reflects the gathering and development characteristics of creative cultural communities in five aspects to the maximum. The utilization rate is large, and the internal space of the site is rich, which can better attract people to enter; commercial space and office space are vertically distributed, business is concentrated on the first and second floors, and office space is concentrated on the 4th to 8th floors to avoid mutual interference, but connected to each other, soitis easy to use. The atrium is introduced to enrich the interior space and increase the landscape interface. The whole project is enclosed, forming a composite interface with the external space environment, limiting the internal spatial scope, and blending with the urban environment.

该项目使用 ideaMaker 软件对 Revit 模型进行处理，使用 Raise3d Pro2 高精度 3D 打印机进行模型三维打印。

使用 3D 打印技术可以提高沟通效率，便于设计师对三维模型的推敲。

通过 BIM 软件（Revit+Lumion），搭建出方案三维模型，可推敲研究设计建筑的外观效果、功能布局、能见度，确保理性、可行性。包括展示及对比不同的项目设计方案、测绘项目四周建筑物及道路的大小及位置、能见度模拟分析等。

This project uses ideaMaker software to process the Revit model and Raise3d Pro2 high-precision 3D printer for 3D printing of the model.

This project is a key one in Chongqing. The use of 3D printing technology can improve communication efficiency and make it easier for designers to modify three-dimensional models.

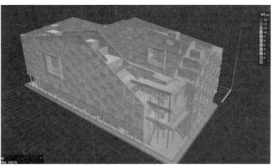

With BIM software (Revit + Lumion), we build a three-dimensional model of the scheme, with which one can study the appearance effect, functional layout and visibility of the designed building so as to ensure rationality and feasibility. It includes displaying and comparing different project design schemes, mapping the size and location of buildings and roads around the project, visibility simulation analysis, etc.

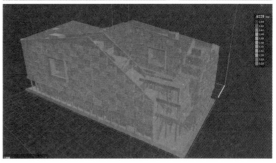

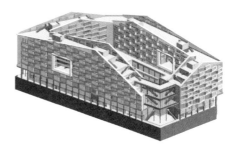

蕴藏的能量
A reservoir of energy

桂阳体育馆
Guiyang Stadium

建筑是在对外部环境做出回应的前提下产生的，在不同时间阶段形成了具有不同特征的建筑。对于城市来说，建筑不应该是一种入侵，也不该去破坏良性的发展秩序和公共空间，而是要去承担起城市使命——我们希望建筑是可以"回应自然"的，是对城市友好的。近几年，友好的城市开放空间这个概念在很多地方被提起，我们希望通过实践为城市创造更多的"开放空间"，桂阳体育馆就是一次这样的探索，以"回应自然"的方式成为城市标志性的区域中心。建筑的设计灵感源自矿石中绽放的宝石晶体。通过使用穿孔铝板的建筑表皮形成律动的起伏，使建筑犹如绿色森林中一组优雅的水晶，建筑坐落在5m高的基座上，有效整合了场馆建筑庞杂的配套设施，化解巨型体育建筑的压迫感，使基座上部的建筑获得轻盈纯净，富于漂浮感的体态特征，创造出和谐的空间环境。使整体建筑形态犹如山体中一组峡谷地形——"凌峻巨石"。

春分

溪边风物已春分。画堂烟雨黄昏。
水沈一缕袅炉薰。尽醉芳尊。

——宋·仲并《画堂春》

"全民健身中心"建筑空间设计研究

—— 以桂阳全民健身中心项目为例

在我国，陆续出现了一座座地标性体育建筑，之所以称之为地标性建筑，是因为它们不仅功能影响面广，意义重要，而且规模较大，造型新颖、奇特，令人震撼。现代体育建筑设计流行态势可总结为：出挑、扭曲、杂乱、无规则。因此在现代体育建筑设计上因过度追求形式而出现了很多误区：形式怪诞、平面不好利用、造价昂贵、维护成本高一味追求复杂性。无规则，并不一定适合我国国情在实践中需要进行技术与经济的综合分析。

1. 项目概况

项目位于桂阳县城市新区欧阳海路延长线，建设在一座缓缓隆起的坡地上，巨型的银灰色建筑以森林绿色和城市建筑为背景，宛如矿山中的宝石晶体，见证桂阳新区的演变，呼应地域特点对场地的影响。场地内主要建筑坐落在 5m 高的基座上，基座有效整合了场馆建筑庞杂的配套设施，化解巨型体育建筑的压迫感，使基座上部的建筑获得轻盈纯净，富于漂浮感的体态特征，创造出和谐的空间环境。基座从东侧、南侧逐步抬升，由硬地与园林小品交织，其图案肌理取自一组峡谷中经过风沙雕琢成形成的巨石，犹如凌峻巨石立于桂阳新区。

2. 设计目标

打造桂阳整个区域的核心地标，营建成功的体育产业，带动周边文化、商业、娱乐、生活休闲和服务产业，形成新型城市综合体，提升桂阳县人们生活整体品质，成为桂阳县体育事业及文化事业发展里程碑。

3. 设计理念

以多种功能有机融合，形成各种活动相互支撑，各种人流相互带动人气，成为富有活力的多样化的城市片区，成为桂阳新时代的体育中心、集会中心，提高城市品位与格调，带动城市消费，产生无形的地标性。

以别具一格的建筑形态形成强烈的可识别性，打造桂阳地标性建筑。建筑设计力求具有地方性、前瞻性，同时又具有实用性、独特感及时代感。合理结合城市文脉与项目自身特征，遵循"以人为本"和"以质为本"的原则。把握未来地标型建筑的发展方向，积极运用高科技设备和细部设计，打造绿色节能建筑的新标杆。高效率的城市地块开发策略，充分挖掘地块的巨大潜力。

4. 建筑设计

全民健身中心为篮球馆、游泳馆及羽毛球馆三部分，其中 A 区篮球馆四周布置固定看台及活动看台，含比赛馆及训练馆，设计时兼顾以后的商业运作，可作为商业演出及大型体育赛事比赛场馆。负一层考虑停车空间。一层、负一层、负二层设置商业空间，以保证以后体育场馆运作。

篮球馆通过上面人行过街通道连接，形成整体。造型现代流畅，建筑群西低东高，逐渐向后面山体过度，形成完美的城市轮廓线。体育馆采用底框桁架结构，形成无遮挡的视觉空间，观众从体育馆里面看场地无压抑感，市民从外面看又可看到体育场馆飘逸整体的造型。

体育馆可容纳约 4300 人，按国家规定为乙级中型体育馆，综合考虑举重馆及其他功能空间。体育馆呈方形，连接游泳馆、举重馆、羽毛球馆、乒乓球馆、与北面体育场结合布置，相辅相成，体育馆占据了园区的大部分空间，处于用地最醒目的位置。二者通过道路及广场相互连接。

建筑通过银灰色穿孔幕墙包裹体育场馆，广场布置当地特有石材，体现建筑力度，与体育场馆相互统一。综合考虑各个体育场馆功能不同要求，银灰色穿孔铝板具有自重轻、施工快等优势，其透光透气的特点还能大大节约赛后的场馆运营成本。银白色穿孔铝板结构内部结构构件在建筑表面形成律动的起伏，建筑整体形象宛如矿山中的宝石晶体。场地内主要建筑坐落在 5m 高的基座上，基座有效整合了场馆建筑庞杂的配套设施，化解巨型体育建筑的压迫感， 使基座上部的建筑获得轻盈纯净，富于漂浮感的体态特征，创造出和谐的空间环境。使整体建筑形态犹如山体中一组峡谷地形——"凌峻巨石"，具有美妙的韵律感和有机的生命感。 在立面上形成跳跃，整体的形象，建筑西低东高，与山体融合。

5. 结语

大型公共体育建筑让我们记住一个城市，也是一个城市的符号。走休闲产业化道路。形成体育中心，形成城市中心综合体，带动产业文化中心，休闲娱乐中心和文化交流中心。体育建筑代表的是一种现象，代表着人们追求健康、幸福、长寿和自我完善的文化现象。体育建筑应与时代融合，与日常生活融合。

体育建筑应该综合各种功能，丰富人们的生活，形成人们生活的中心。

Research on Architectural Space Design of "Nationwide Fitness Center"
——Taking Guiyang Nationwide Fitness Center Project as an Example

In the past decade, landmark sport buildings have appeared around the world, especially in our country. The reason why they are called landmark buildings is that they not only have a wide range of functions and important significance but also enjoy a large scale and a novel, peculiar and shocking appearance. The popular trend of modern sport architectural design is summarized as: overhanging, distorted, messy and irregular. Therefore, there are many misunderstandings in the pursuit of the form in the design of modern sport buildings: a large number of strange forms, unusable planes, expensive construction, high maintenance costs, only pursuing complexity, no rules, not necessarily suitable for national conditions for our country, need to be technical and economical comprehensive analysis.

1. The Project Overview

The project is located on the extension line of Ouyang Hai Road in the new urban area of Guiyang County. It is built on a gently uplifted slope. The huge silver-gray building is set against the background of forest green and urban buildings, like gem crystals in the mine, witnessing the evolution of Guiyang New District and responding to the impact of regional characteristics on the site. The main buildings in the site are located on a 5m-high pedestal. The pedestal effectively integrates the complex supporting facilities of the stadium building, resolves the oppression of the giant sport buildings, and makes the buildings above the pedestal acquire a light, pure, and floating feature, creating a harmonious space environment. The pedestal is gradually raised from the eastern side and the southern side, intertwined with hard ground and miniature gardens.

2. Design goals

The goals are to build the core landmark of the whole area of Guiyang, build a successful sport industry, promote the surrounding culture, commerce, entertainment, life and leisure and service industries, form a new urban complex, improve the overall quality of life of people in Guiyang County, and make it a milestone for the sport industry and cultural development of Guiyang.

3. The Design Concept

With the organic integration of multiple functions, various activities are formed to support one another, and various flows of people will socialize, making a dynamic and diverse urban area, a sport center, and an assembly center in the new era of Guiyang, improving the taste and style of the city, promoting urban consumption, and creating invisible landmarks.

With a unique architectural form, it will create a strong recognizability and become a landmark building in Guiyang. The architectural design strives to be local and forward-looking while being practical, unique, and contemporary. Reasonably combines the urban context and the characteristics of the project itself and follows the principles of "people-oriented" and "quality-oriented". The project grasps the development direction of future landmark buildings, actively use high-tech equipment and detailed design, and creates a new benchmark for green and energy-saving buildings. Efficient urban land development strategy is adopted to fully bring out the huge potential of the land.

4. Architectural Design

The Nationwide Fitness Center consists of three parts: the basketball hall, the swimming hall, and the badminton hall. The basketball hall in Area A is surrounded by fixed stands and activity stands, including the competition hall and training hall. The design takes into account future commercial operations and can be used for commercial performances and as a large-scale sport competition venue. The basement floor is designed as the parking space. Commercial space is set up on the first floor, the first basement floor and the second basement floor to ensure the future operation of the stadium.

The basketball hall is connected by the pedestrian crossing above to form an entity . The external appearance is modern and smooth. The building complex is low in the west and high in the east and gradually goes towards the mountain behind, forming a perfect urban outline. The gymnasium adopts the bottom frame truss structure to form an unobstructed visual space. The audience can see the venue from the inside of the gymnasium without feeling oppressive, and citizens can see the elegant overall shape of the gymnasium from the outside.

The gymnasium can accommodate about 4,300 people. According to the national regulations, it is a medium-sized gymnasium of Class B. The weight-lifting hall and other functional spaces are comprehensively considered. The gymnasium is square, connected to the swimming pool, weight-lifting hall, badminton hall and table-tennis hall. It is combined with the north stadium. The two are connected to each other by roads and squares.

The building wraps the stadium inside through a silver-gray perforated curtain wall, and the plaza is installed with unique local stones, which reflects the strength of the building and is unified with the stadium. Taking into account different functional requirements of various stadiums, the silver-gray perforated aluminum plate has the advantages of light weight and easy construction. The internal structural components of the silver-white perforated aluminum plate structure form rhythmic undulations on the surface of the building, and the overall image of the building is like a gem crystal in a mine. The main buildings in the site are located on a 5m-high pedestal. The pedestal effectively integrates the complex supporting facilities of the stadium building, resolves the oppression of the giant sport building, and makes the building above the pedestal acquire a light pure, and floating feature, to create a harmonious space environment. The overall architectural form is like a group of canyons in the mountain, i.e."steep boulders", with a wonderful sense of rhythm and organic sense of life. On the facade, a leaping and overall image is formed. The building is low in the west and high in the east, blended with the mountain.

5. Conclusion

Large-scale public sport buildings remind us of a city as a symbol of a city. It takes the road of leisure industrialization, forms a sport center, forms a city center complex, promotes the development of an industrial and cultural center, a leisure and entertainment center, and a cultural exchange center. Sports architecture represents a phenomenon, a cultural phenomenon in which people pursue health, happiness, longevity, and self-improvement. Sports architecture should be integrated with the times and daily life. Sport architecture should have various functions, improve people's life and form the center of their life.

桂阳体育馆

Guiyang Stadium

用地面积：
43398m²
建筑面积：
43470m²
容积率：
2.13
设计时间：
2013-2015
建造时间：
2015-2018

桂阳体育馆位于郴州市桂阳县欧阳海路以东，体育馆内共设有篮球馆，游泳馆，羽毛球馆，举重训练馆及乒乓球馆。体育馆呈方形，连接游泳馆、举重训练馆、羽毛球馆、乒乓球馆、与北面体育场结合布置，整体造型现代流畅。建筑群西低东高，逐渐向后面山体过渡，形成完美的城市轮廓线。体育馆采用底框网架结构，形成无遮挡的视觉空间，观众从体育馆里面看场地无压抑感，市民从外面看又可看到体育场馆飘逸整体的造型。

（下图：桂阳体育馆效果图）

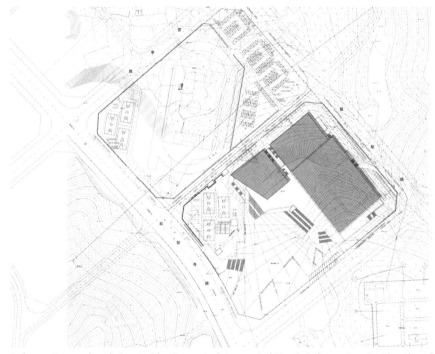

Site Area：
43,398m²
GFA：
43,470m²
FAR：
2.13
Design Time：
2013-2015
Construction Time：
2015-2018

Guiyang Gymnasium is located in the east of Ouyanghai Road, Guiyang County, the City of Chenzhou. There are basketball halls, swimming pools, badminton halls, weight-lifting training halls, and table-tennis halls. The gymnasium is square, connected to the swimming pool, weight-lifting training hall, badminton hall, table-tenn is hall, and the northern stadium. The overall shape is modern and smooth. The buildings are low in the west and high in the east, gradually agreeing with the mountains behind, hence forming a perfect urban outline. The gymnasium adopts the bottom frame grid structure to form an unobstructed visual space. The audience sees the venue from the stadium without feeling oppressive. The public can see the elegant overall shape of the stadium from outside.

（Below: Renderings of Guiyang Stadium）

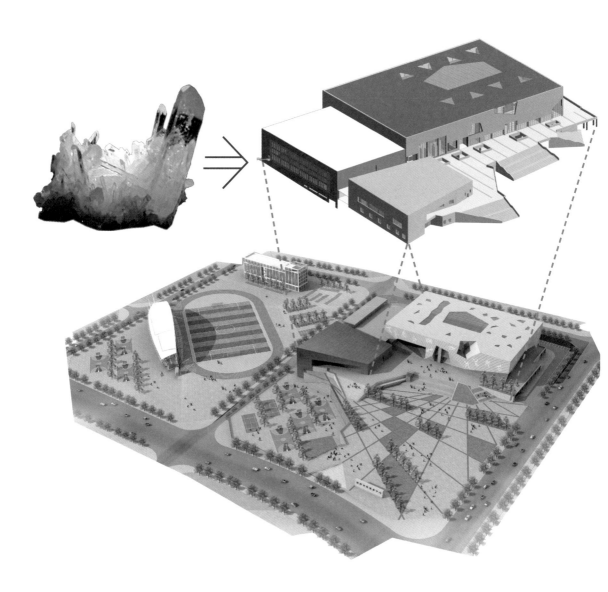

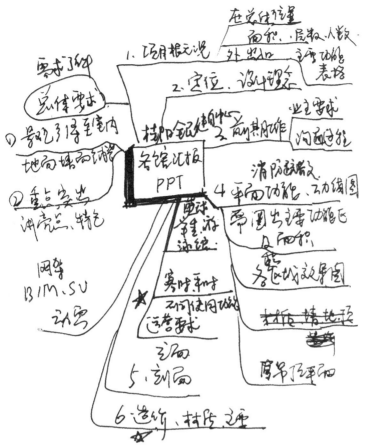

建筑坐落在 5m 高的基座上，有效整合了场馆建筑庞杂的配套设施，化解巨型体育建筑的压迫感，使基座上部的建筑获得轻盈纯净，富于漂浮感的体态特征，创造出和谐的空间环境。使整体建筑形态犹如山体中一组峡谷地形——"凌峻巨石"。

The building is located on a 5m high pedestal which effectively integrates the complex supporting facilities of the stadium building, resolves the oppression of the giant sport building, and makes the building above the pedestal acquire a light, pure and floating feature, to create a harmonious space environment. The overall architectural form is like a group of canyon terrain in the mountain, i.e. "Lingjun boulders".

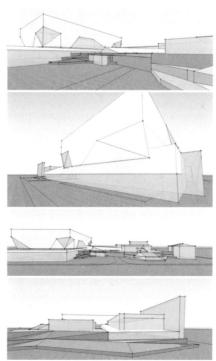

The huge silver-gray building has witnessed the evolution of Guiyang New District against the backdrop of forest green and urban architecture, answering to the regional impact of characteristics on the site. The rhythmic undulated surface of perforated aluminum panels makes the building look like a group of elegant crystals in the green forest.

巨型的银灰色建筑以森林绿色和城市建筑
为背景，见证桂阳新区的演变，呼应地域
特点对场地的影响，通过使用穿孔铝板的
建筑表皮形成律动的起伏，使建筑犹如绿
色森林中一组优雅的水晶。

系统分析图（赛时）

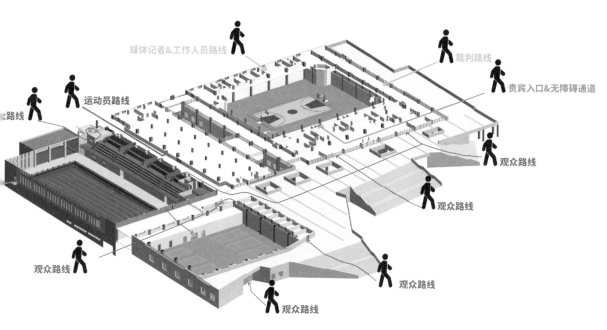

媒体记者&工作人员路线

裁判路线

运动员路线

贵宾入口&无障碍通道

路线

观众路线

观众路线

观众路线

观众路线

观众路线

系统分析图（平时）

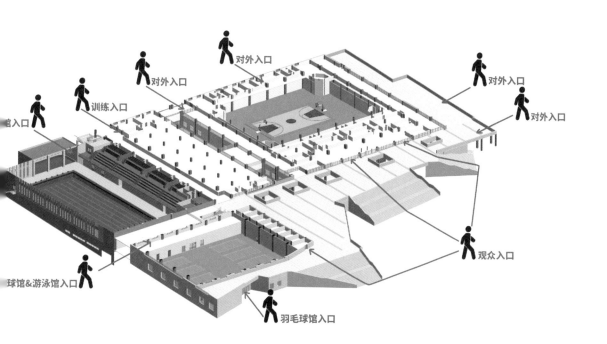

对外入口

对外入口

对外入口

训练入口

对外入口

馆入口

观众入口

球馆&游泳馆入口

羽毛球馆入口

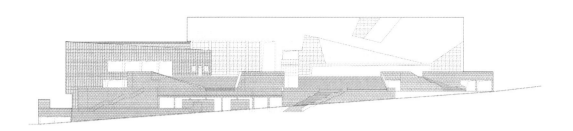

体育中心建筑物包括体育场、体育馆、游泳馆、羽毛球馆、乒乓球馆以及体育艺术走廊配套设施等，银灰色是建筑群统一的外表。穿孔铝板建筑表皮具有自重轻、飘逸等优势，其透光透气的特点还能够大大节约赛后的场馆运营成本。

Sport center buildings include stadium, gymnasium, natatorium, badminton hall, table-tennis hall, sport art corridor, and supporting facilities. Silver gray is the unified appearance of the complex. Perforated aluminum surface building has the advantages of light weight, elegance and so on. Its characteristics of light permeability can greatly reduce the stadium operation cost after the game.

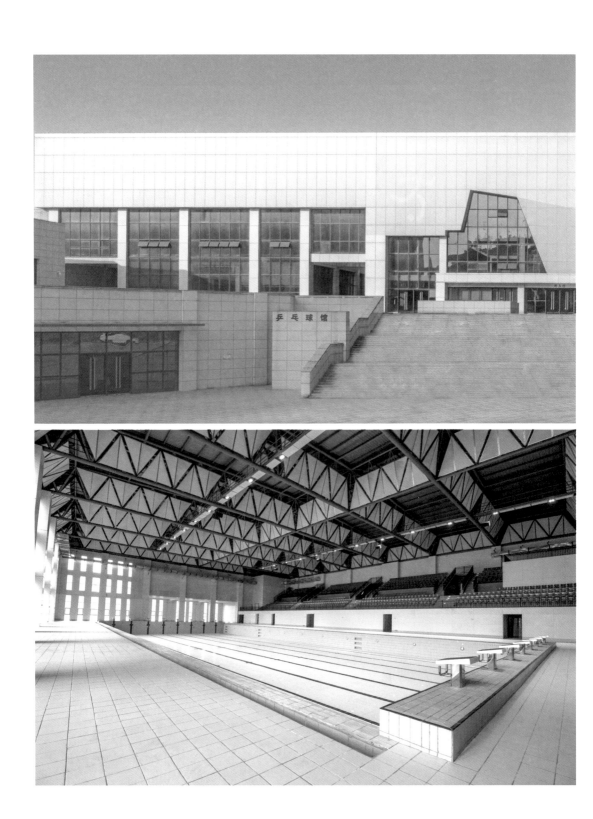

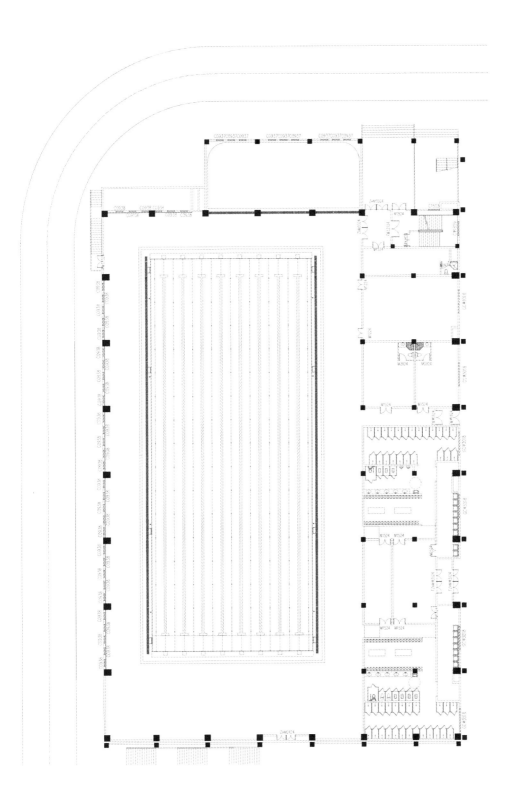

BIM 设计

桂阳县全民健身中心主场馆为 1~3 层（局部带地下室），建筑总长 166.60 m，宽 142.60 m，高 14.40~23.30m。项目的设计过程面临着十分复杂的场地环境，通过 BIM 的碰撞检查和净高验算运用，我们可以站在场景中去推敲空间，推敲材质运用和空间感受。实现对工程施工过程交互式的可视化和信息化管理。

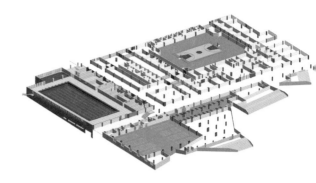

（桂阳体育馆 BIM 剖面图）

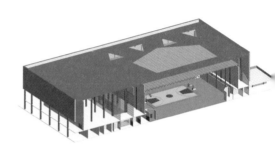

（桂阳体育馆篮球馆 BIM 剖面图）

BIM Design

The Nationwide Fitness Center of Guiyang county consists of 1~3 floors (partially with a basement). The total length of the building is 166. 60 m, the width is 142.60 m, and the height is 14.40~23.30 m. The design process of the project faces a very complex site environment. Through BIM's collision check and net height check application, we can stand in the scene in real time to deliberate on the space, consider the use of materials and the space experience, and fulfill interactive visualization and information management of engineering construction process.

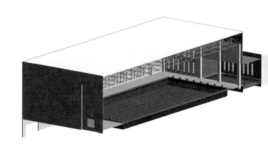

（桂阳体育馆游泳馆 BIM 剖面图）

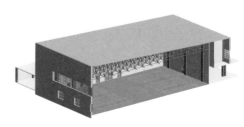

（桂阳体育馆网球馆 BIM 剖面图）

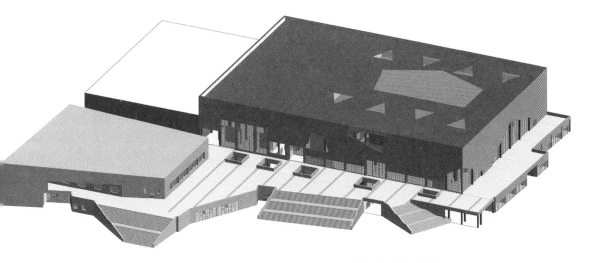

（桂阳体育馆 BIM 鸟瞰图）

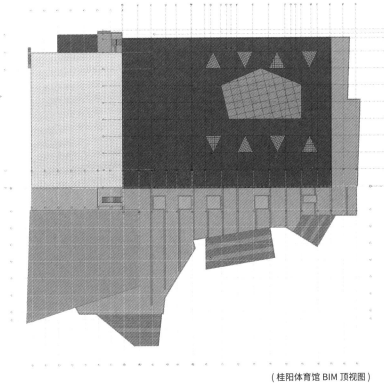

（桂阳体育馆 BIM 顶视图）

聚沙成塔
Gather sand into a tower

湖南创意设计总部大厦
Hunan Creative Design Headquarters Building

长沙是多元文化汇聚之地，我们希望用城市规划和建筑设计来表达和塑造这座城市，一方面，我们在建筑设计中力图展现长沙的创新形象，为日益增多的科技企业提供一个更符合科技企业工作和创新模式的新型办公空间。另一方面，希望以设计引发城市建筑与公共空间的关系和对探索未来长沙城市面貌的思考。设计之初，我们有一个聚沙成塔的概念：项目由商业，基本生活配套及地标塔楼等各种尺度和功能的盒体空间交织而成，像素化的盒子代表着一种包容性，以大小材质不同的像素盒子容纳不同的功能，兼收并蓄的像素值之城也是对湖湘文化开放意识的集中概括。通过空间叠置，使建筑更好地融入城市，并在不影响其他城市环境的基础上更好地利用城市景观资源，将建筑还城于民，并通过垂直绿野的塑造，将自然生态山水意境融入建筑之中，打造自然之丘，森林城市，使二维的空间得到了三维的表达。我们在为长沙创造一个新地标的同时，为项目场地提供一个引人瞩目并令人信服的城市空间。建筑将作为长沙一个新的设计总部标杆，闪耀在马栏山的璀璨天际线上。

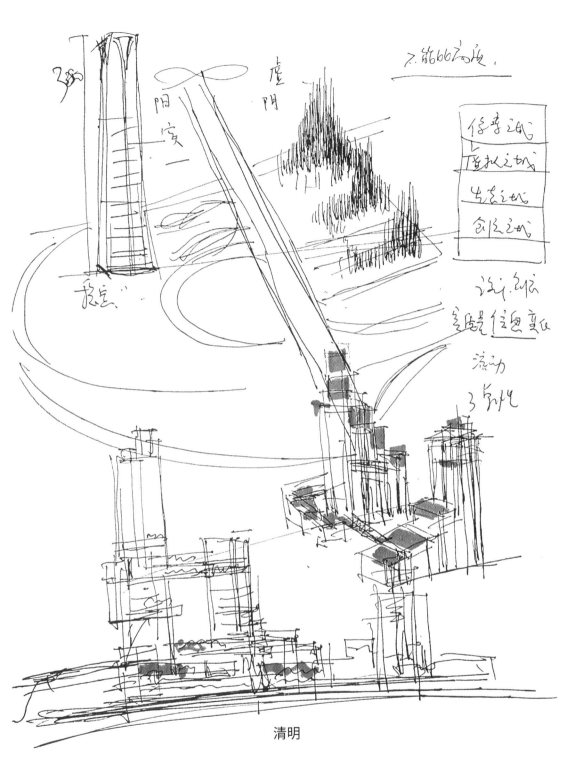

清明

寻游踏青大地春，插枝辟邪柳絮轻。

——清·《二十四节气歌》

夏热冬冷地区绿色低碳建筑研究
—— 湖南创意设计总部大厦

1. 项目概况
（1）项目简介
项目位于长沙马栏山视频文创产业园内，是马栏山文创产业园的第一批启动项目，突出文化创意，定位企业孵化，意打造成城市公园里的文创孵化器和生态公园里的创意孵化基地。总建筑面积 10.29 万 m²，其中地上建筑面积 7.09 万 m²，地下建筑面积 3.20 万 m²。项目由 A、B、C 三栋建筑组成，A 栋为酒店式办公，地上面积 1.26 万 m²，建筑高度 59.55m；B 栋为办公楼，地上面积 3.23 万 m²，建筑高度 99.15m；C 栋为建科院办公楼，地上面积 2.61 万 m²，建筑高度 94.8m。

（2）设计构思
方案提出像素之城、媒体之城、山水洲城的概念。

湖南是多元文化汇聚之地，像素化的盒子代表着一种包容性，以大小材质不同的像素盒子容纳不同的功能。兼收并蓄的像素值之城也是对湖湘文化开放意识的集中概括。每一个像素化的盒子也是一个单元模块，通过灵活多变的组合方式结合装配式建筑的建造方法，打造创意之都。

建筑通过空间叠置，希望能使其更好地融入城市，并在不影响其他城市环境的基础上更好地利用城市景观资源，将建筑还城于民，通过垂直绿野的塑造，使二维的空间得到了三维的表达，增强了内部人员的交流。通过营造由大街、小巷、广场构成的多层级空间，希望与周边环境建立良好的互动，并塑造自身的存在感。建筑根据具体的使用特点将不同的功能空间分解开来，化为相对较小的尺度，旨在寻求或者还原一种适宜的街巷尺度，营造出强烈的社区和城市生活氛围。通过庭院，广场，街巷等空间类型将其组织在一起，从而成为一个建筑群落的聚合体，在其间活动，就像在一个小城市里活动。希望在已经被放大了的城市建筑尺度前提下，仍然能创建一个内在的人性化小尺度空间。

办公楼放弃多余的装饰和堆砌，以简洁明快的体量和纯粹的玻璃幕墙来展现建筑的空间美，结构美。并通过裙房的层层退台及塔楼的体量分解和顶部升起高低错落的造型获得变化，避免单调感，体现该建筑独特的形象和现代特征。

多层次的景观布置使景观空间相互渗透、连续流畅，情趣变化，使高层办公都能享受到多维度的景观。层层跌落的退台景观与建筑空间的结合设置，使建筑不用脱离在自然条件之外，而是与自然融为一体。高低起伏的天际线不仅缓解了对城市的压抑感，而且也使得立体景观特点鲜明，并与南侧城市公园绿轴相呼应。

（3）定位
创全优工程、省优秀设计、争创鲁班奖。
建设过程中，严格秉承"一流、超越、精作、奉献"的企业精神，着力打造绿色建筑、装配式建筑、智慧建筑、BIM 全过程应用等示范。

2. 主要技术措施
（1）绿色建筑
该项目是贯穿设计、施工、运营的全过程综合性绿色建筑示范项目，其中 A、B 栋为二星级绿色建筑，C 栋为三星级绿色建筑。

（2）装配式建筑
项目运用多种装配式技术：A 栋为 PC 装配式结构，装配率 71.2%，同时采用共轴承插型装配式卫生间；B 栋为现浇混凝土 + 钢结构相结合的混合结构，装配率 58.9%；C 栋为钢结构，装配率 84%。

装配式钢结构具有抗震性能好，施工方便，工期短，建筑使用率高，降低主体结构和基础工程成本等优点。

（3）智慧建筑

充分构建融合系统，全方位智能化管控建筑运行和维护，用户充分感受便捷、信息化和科技化的真实体验，助力物业管理和运行维护高效、安全、智能、便捷，力争成为湖南及全国智慧建筑代表性示范工程。主要包括楼宇基础信息化系统、智慧会议室展厅租赁运营系统、V阅平台运营服务系统、智能用电数字化精准运行系统、梯控及智慧通行系统、云中枢平台数字孪生建筑等六大核心内容。其中楼宇基础信息化系统包含楼宇管理、安防管理、生活管理、办公管理及信息基础设施；智慧会议室展厅租赁运营系统园区共享，可实现自助预订、自动计费、服务配套、商业互通，支持高清会议和远程会议等；V阅平台运营服务系统通过楼内显示设备构建展示平台，客户可以利用展示平台进行广告发布，以共享展示为载体，提供视频、广告制作相应服务；梯控及智慧通行系统可提高通行效率、智能引导停车，智能访客系统，提高访客尊贵感受；人脸识别闸机，进出无人值守管理；电梯智控联动，减少等候电梯时间。

（4）BIM 技术应用

通过 BIM 对项目建设全过程进行专业数据集成，数据迭代，实现项目数据完整性、专业数据协调性、数据传递一致性。结合施工过程数据的模型更新，形成可指导项目运营的完整数据模型，从而真正实现基于 BIM 的全过程管理（一模到底）。

贯彻"前策划、后评估"管理思想，项目前期，参建各方充分沟通，避免后期大规模变更等资源浪费，实时对当前阶段的工作成果及时进行评价，并为下一步的策划提供反馈。同时，BIM 与绿建、装配式、装饰装修等各专业配合，打造全专业 BIM 模型展示及模型应用（整体漫游）。

（5）园林景观及海绵城市

场区采用下凹式绿地、透水铺装、植草砖车位、雨水收集回用等海绵城市技术，年径流总量控制率达 75% 以上。

打造包括架空层、屋顶花园、阳台、天桥、墙面等全方位立体绿化。采用装配式绿化技术，利用新技术新材料，通过提前选苗育苗形成装配式种植块或种植槽，现场快速安装。其特点是材料材质轻、施工周期短、季节因素影响小、施工管理和后期维护费用低、水肥管理智能化、可视化。

3. 结语

该项目在设计和建设过程中深入融合"创新、协调、绿色、开放、共享"理念，将成为湖南省乃至全国建筑新技术集成典范，通过该项目培养一批设计、施工、运营及管理人才队伍，助力集团下属建科院、德顺电子、五公司等相关本版块业务的发展，为集团打造高端商业地产品牌，为园区提供高档办公孵化基地。

Research on Green Low-carbon Buildings in Hot Summer and Cold Winter Area
——Creative Design Headquarters Building, Hunan Province

1. The Project Overview

(1) The Introduction

This project is located in Malanshan Cultural and Creative Industrial Park as the first start-up project, highlighting cultural creativity and positioning business incubation, which is intended to be a cultural and creative incubator in the city park and a creative incubation base in the ecological park. The total construction area is 102,900m², of which 70,900m² is above ground and 32,000m² is under the ground. The project consists of three buildings, A, i.e. B and C. Building A is a hotel-style office with an above-ground area of 12,600 square meters and a building height of 59.55m; Building B is an office building with an above-ground area of 32,300 square meters and a building height of 99.15m. Building C is an office building of the Academy of Construction Science with an above-ground area of 26,100 square meters and a building height of 94.8m.

(2) The Design Concept

The program proposes the concepts of a City of Pixels, a City of Media and a City of Mountain and Water.

Hu'nan Province is a place where multiple cultures converge, and the pixelated box represents a kind of inclusiveness, accommodating different functions with pixel boxes of different sizes and materials. The city of pixels is also a concentrated summary of the openness of Hu'nan culture. Each pixelated box is also a unit module, which is combined with the construction method of assembly buildings through flexible and versatile combinations to form a creative city.

By superimposing the space, we hopes to make the building better integrated into the city and make better use of the urban landscape resources without affecting other urban environments, thus returning the building to the citizens. Through the shaping of the vertical green field, the two-dimensional space is expressed in three dimensions, enhancing internal communication among people. By creating a multi-layered space consisting of streets, alleys and squares, it is hoped to establish good interaction with the surrounding environment and shape its own sense of existence. The building decomposes different functional spaces into relatively small scales according to specific characteristics of use, aiming to seek or restore an appropriate street scale and create a strong atmosphere of community and urban life. It is organized together through courtyards, squares, streets, alleys, and other spatial types, thus constructing an aggregation of architectural clusters where people would socialize as if being in a small city. It is hoped that an inherently humane small-scale space can be created despite the already magnified scale of urban architecture.

The office building abandons superfluous decoration and stacking, and uses simple and bright volume and pure glass curtain wall to show the spatial grace and structural grandeur. Through the layer-by-layer retreat of the podium and the decomposition of the tower-volume and the shape of the top rising high and low to produce changes, it is to avoid the sense of monotony and present the unique image and modern characteristics of the building.

The multi-layered landscape arrangement makes the landscape space interpenetrating, continuous and smooth, and interestingly changing so that all the high-rise offices can enjoy the multi-dimensional landscape. The combination of the landscape and building space is set up by the falling layers of the retreat so that the building does not need to be separated from natural conditions, but integrated with Nature. The undulating skyline not only relieves the depression to the city but also makes the three-dimensional landscape distinctive, agreeing with the green axis of the city park on the southern side.

(3) Orientation

The project will be a total excellent project, provincial excellent design, striving for the Luban Award. During the construction process, the company strictly adheres to the enterprise spirit of "first-class, improvement, fine work dedication", striving to build green building, assembly-type building, intelligent building, BIM full process application, and other demonstrations.

2. Main technical measures

(1) Green Buildings

The project is of comprehensive green building demonstration throughout the whole process of design, construction, and operation, of which Building A and B are two-star green buildings and Building C is a three-star green building.

(2) Assembled buildings

The project uses various assembly technologies, i.e. Building A adopts a PC assembly structure with an assembly rate of 71.2% and a common bearing insert assembly bathroom; Building B is a mixed structure combining cast-in-place concrete and steel structure with an assembly rate of 58.9%; Building C is a steel structure with an assembly rate of 84%.

The assembled steel structure has the advantages of good seismic performance, convenient construction, a short construction period, a high building utilization rate, and a reduced cost of main structure and foundation works.

(3) Intelligent buildings

Efforts are made to fully build an integrated system, all-round intelligent control of building operation and maintenance, so that users would fully feel the convenience, information technology and technology-based real experience, help with property management as well as efficient, safe, intelligent and convenient operation and maintenance, striving to become a representative demonstration project of intelligent building in Hu'nan Province and China . It mainly includes six core contents of building foundation information system, intelligent conference room exhibition hall leasing operation system, V reading platform operation service system, intelligent electricity digital operation system, ladder control and intelligent traffic system, and cloud hub platform digital twin building. Among them, the building basic information system includes building management, security management, life management, office management, and information infrastructure; the intelligent conference room and exhibition hall rental operation system is shared throughout the park, which can realize self-reservation, automatic billing, service support, and commercial interoperability while also support high-definition conference and remote conference. The V reading platform operation service system offers a display platform through the display in the building, and customers can use the platform for advertisement to provide services such as video and advertisement customization via the platform. The ladder control and intelligent traffic system can improve the traffic efficiency, intelligent parking, intelligent visitor system, and the visitor feeling of nobility. The face recognition gate enables man-free gate management. The intelligent elevator control reduces the time that one has to wait for the elevator.

(4) BIM technology application

Professional data integration and data iteration through BIM for the whole process of project construction to achieve project data integrity, professional data coordination and data transmission consistency. Combined with the model update of the construction process data, a complete data model can be formed to guide the project operation, so as to truly realize the whole process management based on BIM (a model for full process).

Implementing the management idea of "pre-planning and post-evaluation". In the early stage of the project, all parties concerned in the project communicate with one another to avoid the waste of resources such as large-scale changes in the later stage, evaluate the work results of the current stage in real time, and provide feedback for the next planning. At the same time, BIM and green construction, assembly, decoration and other professional cooperation serve to create a professional BIM model display and model application (overall roaming).

(5) Garden landscape and sponge city

The site adopts sponge city technology such as recessed green space, permeable pavement, grass tile parking space, rainwater collection and reuse, with the total annual runoff control rate reaching more than 75%.

Create a full range of three-dimensional greening including elevated floors, roof gardens, balconies, catwalks, walls, etc. The use of assembled greening technology, the new technology, and new materials, through the early selection of seedlings, serve to form assembled planting blocks or planting troughs, on-site rapid installation. It is characterized by light materials, a short construction period, low influence of seasonal factors, low construction management and post-maintenance costs, intelligent and visualized water and fertilizer management.

3. Conclusion

With the concept of "innovation, coordination, green and openness" during the design and construction process, this project will become a model of new construction technology integration in our province and even in China. Through this project, a group of design, construction, operation and management talents will be cultivated, which will help the Group's subsidiaries, such as Academy of Construction Sciences, Desun Electronics, and Five Companies to develop their business in this section, and create a high-end commercial real estate brand for the Group, hence providing a high-grade office incubation base for the park.

湖南创意设计总部大厦

Hunan Creative Design Headquarters Building

用地面积：
29092m²
建筑面积：
94440.5m²
容积率：
3.5
设计时间：
2019-2020
建造时间：
2020-2022

湖南创意设计总部大厦选址于长沙马栏山视频文创产业园内地块 18（X06-A49）、地块 19（X06-A56-1），位于东二环以东，鸭子铺路以北，滨河路以南，滨河联络路以西，区域位置极佳。项目总用地面积为 29092 m²。建筑通过空间叠置，希望能使其更好地融入城市，并在不影响其他城市环境的基础上更好地利用城市景观资源，将建筑还城于民，通过垂直绿野的塑造，使二维的空间得到了三维的表达，增强了内部人员的交流。

（下图：湖南创意设计总部大厦总图）

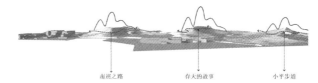

南巡之路　　　　春天的故事　　　　小平步道

Site Area：
29,092m²
GFA：
94,440.5m²
FAR：
3.5
Design Time：
2019-2020
Construction Time：
2020-2022

Creative Design Headquarters of Hunan Province is located in plot 18 (X06-A49) and plot 19 (X06-A56-1) in Malanshan Video Cultural and Creative Industry Park, Changsha, east of East Second Ring Road, north of Yazipu Road, south of Binhe Road, and west of Binhe Lianluo Road. The area is very well located. The total land area of the project is 29,092m². By superimposing the space, the building is hoped to make the building better integrated into the city and make better use of the urban landscape resources without affecting other urban environments, hence returning the building to the citizens. Through the shaping of the vertical green field, the two-dimensional space is expressed in three dimensions, thus enhancing internal communication among people.

(Below:Creative Design Headquarters Building of Hunan Province)

2019 年 5 月 28 日在集团
三公司设计院、省建科院
共征集到 9 个方案进行比
选，经评选推荐 6 个方案
于 2019 年 6 月 9 日报长
沙市政府，确定省建科院
设计六所的方案为实施方
案，并将该方案报省委省
政府主要领导审查通过。

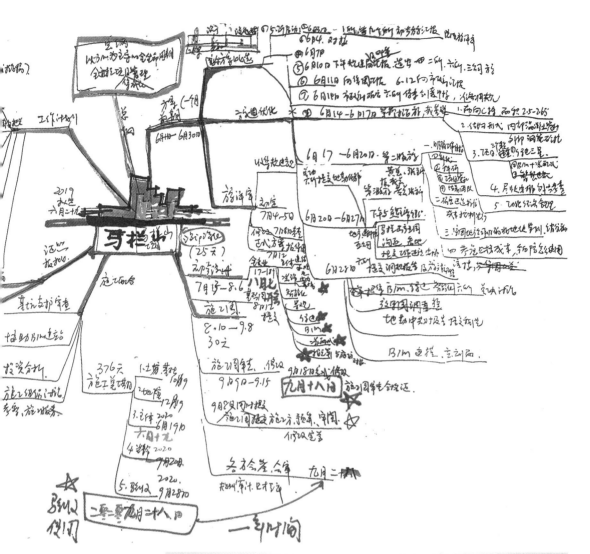

On May 28, 2019, a total of 9 schemes were collected from the design institute of three companies of the group and the Provincial Academy of Construction Sciences for comparison and selection. The plan of the sixth Institutes of the Academy of Construction Science is selected to be the implementation plan, and the plan is reported to the main leaders of the provincial Party committee and government for review and approval.

The office building abandons superfluous decoration and stacking, whereas usingsimple and bright volume and pure glass curtain wall to show the spatial grace and structural grandeur. Through the layer-by-layer retreat of the podium and the decomposition of the tower-volume and the shape of the top rising high and low to produce changes, to avoid the sense of monotony, hence presenting the unique image and modern characteristics of the building. It aims to create cultural incubators in city parks and creative incubation bases in ecological parks.

办公楼放弃多余的装饰和堆砌，以简洁明快的体量和纯玻璃幕墙来展现建筑的空间美，结构美。并通过裙房的层层退台及塔楼的体量分解和顶部升起高低错落的造型获得变化，避免单调感，体现该建筑独特的形象和现代特征。打造城市公园里的文化孵化器，生态公园里的创意孵化基地。

该项目由 A 栋 16 层酒店，B 栋 22 层高层办公，C 栋 21 层办公楼，整体 2 层地库组成。A 栋 16 层，一、二层层高 4.5m，四～十六层层高 3.6m，建筑高度 59.55m。B 栋办公楼层高 4.5m，建筑高度 99.15m。C 栋 21 层办公楼，层高 4.5m，建筑高度 94.8m。

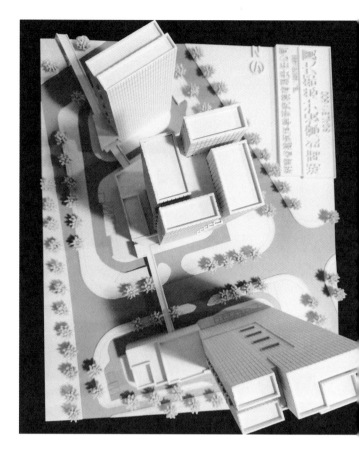

The project consists of a 16-story hotel of Building A, a 22-story high-rise office of Building B, a 21-story office building of Building C, and an overall 2-story basement. Building A has 16 floors, the first and second floors are 4.5m high, the fourth to sixteen floors are 3.6m high, and the building height is 59.55m. The building B office floor is 4. 5m high and the building height is 99.15m. The 21-story office building C is designed with a floor height of 4.5m and the total building height is 94.8m.

空间的交汇
Intersection of Spaces

创元时代
Chuangyuan Times

场地是长沙市地铁 5 号线与地铁 8 号线（规划）的交汇处，且与规划中的地铁 5 号线出口相接驳，天然的"交汇"属性毋庸置疑的成了我们设计的出发点与中心点。"我们相信，与自然环境更近的距离可以提高人脑的活跃度，可以激发创意产业者的思路。"Eric Philips 对《好奇心日报》说。因此我们希望创造一个融入周边环境的建筑体，在现状和规划用地之间，穿越万家丽高架，及未来规划的湘府路高架，规划建设一个复合型社区，紧密组织高架四周社区，起到"聚合"作用。强调从街道到建筑空间的无缝衔接，对于在此工作的每一个人，生活不会戛然而止，而工作时间随时开启。虽然建筑形态呈圆形，但总体布局推演过程严格遵循商业逻辑。建筑形态经过被挤压、揉搓、捏合，得以形成入口、阳台、特殊的共享空中屋顶花园。我们试图模拟街道生活、复制城市功能，希望能将建筑空间打造出社区感，加强人与人的连接，人与城市的连接。

谷雨

二月山家谷雨天，半坡芳茗露华鲜。

——唐·陆希声《阳羡杂咏十九首·茗坡》

城市复合型社区的地产开发模式研究
—— 以创元时代项目为例

随着城镇化进程加快，人们生活水平不断提高，人们不再仅仅满足于住宅所提供的居住功能，原有的住宅区建设也就有了新的发展。住宅区作为城市的有机部分，也需要承担部分城市的服务功能。为满足居民日益多样化的生活需求以及发挥城市服务功能的需要，住宅区正在向房地产与其他关联产业为一体的复合型地产进行转变。因此，城市复合型社区应运而生。复合型社区区别传统的单一功能住宅区，根本特征在于其具有整合性，能够将不同形式、不同功能的产业有机聚合在同一地产上，从而增加地产的经济效益，是一种地块功能与城市功能相叠加，内部居住与城市服务功能相结合的地产结构。

1. 项目概况

创元时代项目位于长沙市雨花区，湘府路与万家丽路西北角。地理位置优越，周边基础设施完善，周边不仅有大型商业综合体喜盈门·范城、华悦城、德思勤城市广场等，该项目还有丰富的景观资源，往西边步行 5 分钟便是被称为"城市绿肺""天然氧吧"的圭塘河风光带，往南边步行 10 分钟便是国家 4A 级旅游景点的湖南省森林植物园。项目用地北侧为烟草公司，西侧为托斯卡纳项目，用地东侧为万家丽高架桥、美洲故事住宅小区，南侧为湘府路高架。是地铁 5 号线与规划地铁 8 号线的交汇处，与规划中的地铁 5 号线出口相接驳，交通便捷这不仅缩短了市民的生活圈，还能带动项目的商业活力。项目总规划用地为约 58 亩左右，该项目基地因在航空线限高范围内，建筑高度得以控制在 100m。

2. 设计理念

该项目是一个属于公众的开放性城市空间。规划设计首先必须尊重周边环境，与城市融为一体，被环境所接收才能充分汇聚人气。因此设计了连接现实与艺术的空中连廊，打造了聚集人气与未来的商业广场。

空中走廊——连接到各个商业功能区的主要"动脉"，为商业流线梳理起到了决定性的作用。它也将会迎来更多的商业公共空间，广场为城市活动和艺术表演提供场所。设围合商业广场，营造商业氛围，给消费者带来自然的购物体验。地面超大广场同时接驳城市地铁空间，带来丰富的消费人群，也缓解地铁人流的疏散压力。

3. 规划设计

（1）思未来

在对原始建筑体块进行多次切割推演之后，形成了最后宛如城市宝石一般的建筑造型，就像顽石经过千雕万琢方能形成璀璨宝石一般，我们的建筑设计灵感也来源于此。项目汇聚都会、生态、商务、商业、娱乐文化等元素，集高端住宅、甲级写字楼、风情商业街等多种形态于一体，占地约 36 亩，总建筑面积约 11 万 m^2。由 1 栋 19 层写字楼、3 栋高层住宅、1 栋商业裙楼组成。

（2）运新技

通过建立 BIM 模型，对方案进行模拟分析，提前预测到各个方案的可行性，综合考虑优选出最佳方案。配合设计工作，建立全专业的 BIM 模型，基于 BIM 模型生成高质量的设计施工图纸。通过全专业的协同工作，实现一处修改，处处修改。

4. 结语

创元时代造型美观新颖，并且通过严密的分析，在平面布局，天际线变化等城市设计层面上充分考虑到城市未来的发展，希望通过该项目，结合长沙南部的发展，为长沙南部的未来城市形象做出贡献，成为南部的一个有代表性的城市复合型社区以及区域级城市标志性建筑，打造南部一块绚丽的蓝宝石！

Research on the Real Estate Development Model of Urban Compound Community
——Taking the Project of Changyuan Times as an Example

With the acceleration of urbanization and the continuous improvement of people's living standards, people are no longer satisfied with the living functions provided by houses, and the construction of the original residential areas has also undergone new development. As an organic part of the city, the residential area also needs to undertake some service functions of the city. In order to meet the increasingly diversified living needs of residents and the need to implement urban service functions, residential areas are being transformed into complex real estate which integrates real estate and other related industries. Therefore, urban complex communities emerge as required by the time. Different from the traditional single-function residential area, the compound community is fundamentally characterized by its integration, which can organically aggregate industries of different forms and functions on the same real estate, thereby increasing the economic benefits of the real estate. Urban functions are superimposed, and the real estate structure is a combination of internal residence and urban service functions.

1. The Project Overview

The project of Chuangyuan Times is located in Yuhua District, the City of Changsha, at the northwest corner of Xiangfu Road and Wanjiali Road. The geographical position is superior, and the surrounding infrastructure is complete. Not only are there large commercial complexes such as CIMEN·FUN city, JOY city, TASKIN city plaza, etc., the project also has rich landscape resources. The Guitang River Scenic Belt, known as "the urban green lung" and "natural oxygen bar", is only a 5-minute walk to the west. Hu'nan Forest Botanical Garden, a national 4A-level tourist attraction, is only 10-minute walk to the south. The northern side of the project site is the Tobacco Company, the western side is the Tuscany Project, the eastern side of the site is Wanjiali Overpass and the American Story Residential Community, and the southern side is the Xiangfu Road Overpass. It is the intersection of Metro Line 5 and the planned Metro Line 8, connected to the planned exit of Metro Line 5. The transportation is convenient, which not only shortens the living circle of citizens but also promotes the commercial vitality of the project. The total planned land for the project is about 58 mu. The project base is within the height limit of the aviation line, and the building height can be controlled at 100 meters.

2. The Design Concept

This project is an open urban space belonging to the public. Planning and design must first respect the surrounding environment, integrate with the city, and be accepted by the environment in order to attract traffic. Therefore, an air corridor connecting reality and art was designed to create a commercial plaza for its popularity and the future.

The air corridor, the main "artery" connecting to various commercial functional areas, plays a decisive role in sorting out the commercial streamline. It will also welcome more commercial public spaces, with plazas providing venues for urban activities and artistic performances. Set up an enclosed commercial plaza to create a commercial atmosphere and offer consumers a natural shopping experience. The super-large square on the ground is also connected to the urban subway space, bringing in lots of consumer groups and relieving the evacuation pressure of subway passengers.

3. Planning and Design

(1) Thinking about the future

After cutting and deduction of the original building block several times, the final architectural shape like an urban gem was formed, just like a sturdy stone which can be formed into a splendid gem after thousands of carvings, Our architectural design inspiration also comes from this. The project brings together elements of metropolis, ecology, business, commerce, entertainment and culture, integrating various forms such as high-end residences, grade-A office buildings, and stylish commercial streets. It covers an area of about 36 acres and has a total construction area of about 110,000 square meters. It consists of one 19-story office building, three high-rise residential buildings, and one commercial podium building.

(2) Application of New Technology

By establishing a BIM model, simulating and analyzing the scheme, predict the feasibility of each scheme in advance, and pick the optimal scheme after comprehensive consideration. Cooperate with the design work, establish a professional BIM model, and generate high-quality design and construction drawings based on the BIM model. Through the professional collaborative work, one can modify the model place by place.

4. Conclusion

Chuangyuan Times has a beautiful and novel shape. Through rigorous analysis, the future development of the city is fully considered at the urban design level such as the plane-layout and skyline changes. It is hoped that this project, in line with the development of southern Changsha, will contribute to the image of the future city in southern Changsha, and become a representative urban complex community in the south, a landmark building in a regional city, and a gorgeous sapphire in the south!

创元时代

Chuangyuan era

用地面积：
39165.7m²
建筑面积：
108574.8m²
容积率：
3.51
设计时间：
2016-2017
建造时间：
2017-2019

成兴创元时代项目位于长沙市雨花区，湘府路与万家丽路西北角。该项目是地铁 5 号线与地铁 8 号线（规划）的交汇处，交通区位优势明显；且项目与规划中的地铁 5 号线出口相接驳，出行较为便利。项目总用地面积 39165.7m²，净用地面积 23704.1m²，综合容积率 3.51，总建筑面积约 10.7 万 m²。人防必建面积 3433m²，设二个甲 6 级人员掩蔽所，项目定位为由高层住宅、高层商务楼、商业构成的城市小型综合体。

（下图：创元时代总图）

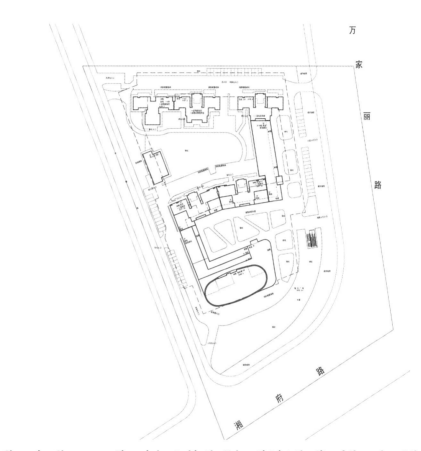

Site Area：
39,165.7m²
GFA：
108,574.8m²
FAR：
3.51
Design Time：
2016-2017
Construction Time：
2017-2019

Chengxing Chuangyuan Times is located in the Yuhua District, the City of Changsha, at the northwest corner of Xiangfu Road and Wanjiali Road. The project is at the intersection of Metro Line 5 and Metro Line 8 (planning), which means transportation convenience. The total land area of the project is 39,165.7 m², the net land area is 23,704.1m², the comprehensive plot ratio is 3.51, and the total construction area is about 107, 000m². The area of civil air defense that shall be built is 3,343m². There are two level A6 personnel shelters. The project is positioned as a small urban complex consisting of high-rise residential buildings, high-rise office buildings and commercial buildings.

(Blowe: General map of Chuangyuan Times)

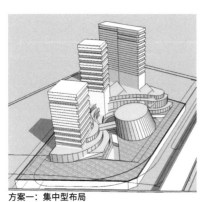

方案一：集中型布局

容积率 FAR=3.15

方案二：集中＋散落式布局

容积率 FAR=2.60

方案三：内庭院式布局

容积率 FAR=3.90

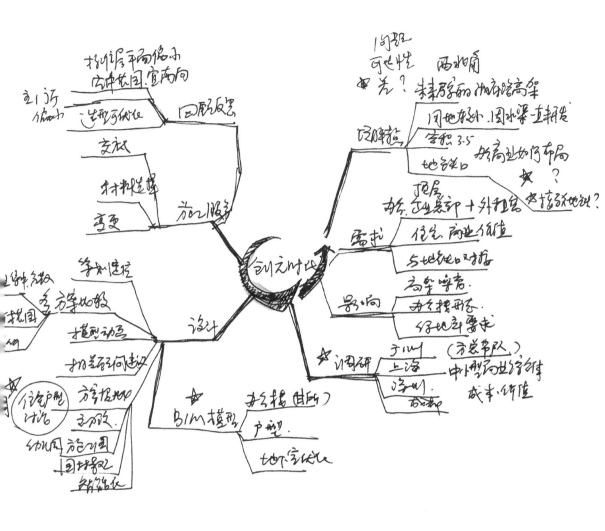

方案四：内街式布局
容积率 FAR=3.28

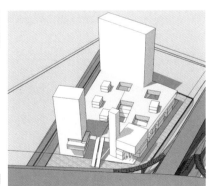

方案五：满铺式布局
容积率 FAR=2.80

方案六：复合通透式布局
容积率 FAR=3.49

该项目总用地面积 40097.1 m²，净用地面积 23699.4 m²，综合容积率 3.49，总建筑面积约 107129.7 m²。项目定位为由高层住宅、高层商务楼、商业构成的城市小型综合体。

都市山水

远离喧嚣城市，置身于公园魔方，给心灵一片安逸和清新回归，对都市人有魔一般的吸引力。

创客空间

在现状和规划用地之间，穿越万家丽高架，及未来规划的湘府路高架，规划建设一个复合型社区，紧密组织高架四周社区，起到"聚合"作用。

The total land area of this project is 40,097.1m² , the net land area is 23,699.4m² , the comprehensive plot ratio is 3. 49, and the total construction area is about 107,129.7m². The project is positioned as a small urban complex consisting of high-rise residential buildings, high-rise office buildings, and commercial buildings.

Urban landscape

Away from the hustle and bustle of the city, being in the rubik's cube in the park gives the soul a sense of ease and freshness, and reveals a magical appeal to urbanites.

Maker Space

Between the status quo and planned land, through the Wanjiali Overpass and the future planned Xiangfu Road Overpass, a complex community is planned to be built. The surrounding communities by the overpass are closely organized to play the role of "aggregation".

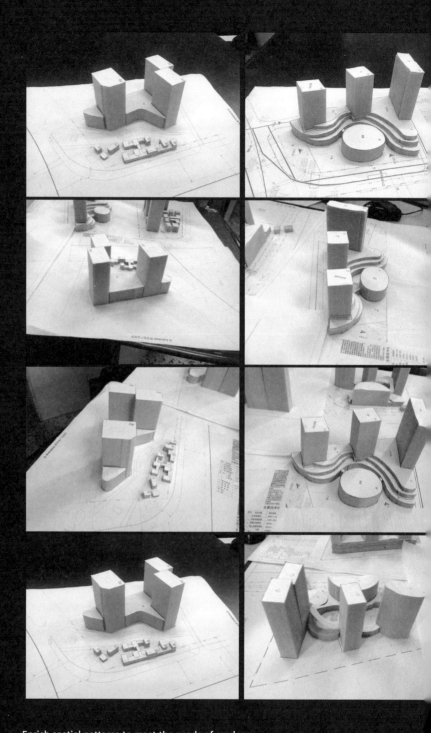

Enrich spatial patterns to meet the needs of modern business formats and promote commercial benefits. Ecological, experiential, open, composite functional and combined blocks.

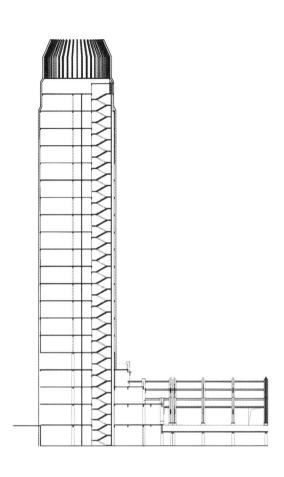

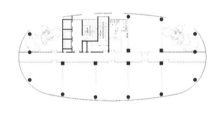

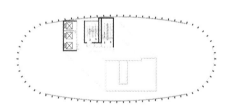

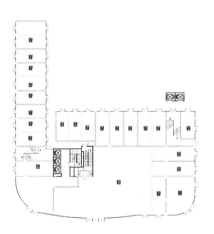

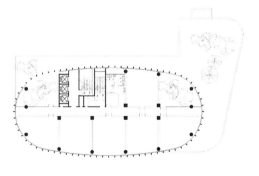

生长与聚落
Growth and settlement

山南市隆子县玉麦乡玉麦小康村
The Construction Project of Yumai Xiaokang Village, Yumai Township, Longzi County, the City of Shannan

格桑花是高原幸福和爱情的象征，也是藏族人民心中永远的追求。过去，玉麦这个位于西藏自治区山南市3600m 常年大雪封山半年以上的"孤岛"鲜为人知，这里距离县城 190km，需要翻过两座 4000m 的雪山、一座 5300m 的雪山才能抵达，山路崎岖，陡峭险峻，其中包括 180km 以上的盘山临崖土路。这里流传着一个几代人默默无闻、扎根祖国边陲放牧守边的故事；几代人甘于奉献、捍卫 3644km² 土地上一草一木的足迹；几代人执着坚守、用半个世纪凝铸"家是玉麦，国是中国"的壮举，就发生在茫茫雪山怀抱中。党的十九大闭幕后，习近平总书记给西藏隆子县玉麦乡牧民卓嘎、央宗姐妹回信。信中对父女两代人在边境高原上默默守护祖国领土，表示"崇高的敬意和衷心的感谢"。

玉麦乡位于喜马拉雅山北支脉日啦山、扎日山、博沙拉山南麓，喜马拉雅山主脉北麓的中印边境藏南地区，面对独特的自然环境，我们试图在改变和保留中寻求微妙的平衡，衡量玉麦空间在呈现后的长周期中的经营可持续性，是否在应对外界变化中展现出空间的韧性、可变性与生长潜力。以独特的空间承载更具深远影响的文化形态，使空间成为穿越时间的文化元素，为地区文化的发展提供持续不断的养分。家是玉麦，国是中国，放牧守边不仅是她们的职责，也是我们的职责。年轻一代玉麦人，传承和发扬前辈们的精神，当好神圣国土的异乡守护者、守边固边兴边。

立夏

晴日暖风生麦气，绿阴幽草胜花时。

——宋·王安石《初夏即事》

乡村振兴战略背景下民族地区的乡村建设

—— 山南市隆子县玉麦乡玉麦小康村建设

实施乡村振兴战略、区域协调发展战略，是党的十九大提出的建设现代化经济体系的重大任务。坚持总体国家安全观，大力加强国防建设，维护边境安全，是新时代坚持和发展中国特色社会主义的重大基本方略。加快西藏边境地区小康乡建设，促进农村贫困人口加快脱贫，实现村民与全国同步全面建成小康社会的宏伟目标，对与西藏实现"两个一百年"奋斗目标具有重要意义，其经济效益、社会效益和政治效益十分显著。随着我国全面建成小康社会进入决胜阶段，"一带一路"建设加快推进，区域协调发展不断深化，脱贫攻坚全面展开，国家对边境地区全方位扶持力度不断加大，我国与周边国家关系的发展进入新阶段，深入推进兴边富民行动面临难得的机遇。

1．项目概况

玉麦乡位于喜马拉雅山北支脉日啦山、扎日山、博沙拉山南麓，喜马拉雅山主脉北麓的中印边境藏南地区。玉麦乡村驻地平均海拔3560m，印度洋暖湿气流沿玉麦河谷北上，遇到海拔5000m的日啦山阻挡，造成玉麦冬季多雪、春夏秋三季多雨的气候特点，每年11月至来年5月是大雪封山时期。项目总用地面积29.40公顷，总建筑面积17273.25m²，项目总投资约1.1亿元。项目建成后惠及本地居民56户192人。其中一期建设工程包含56栋民居，1栋多层乡政府、村委会，1栋多层小学、幼儿园，1栋多层卫生院，1栋多层陈列馆及两栋公厕，一期建设总建筑面积为11612.79m²。根据西藏自治区对项目建设的总体要求，考虑到玉麦乡多雨、寒冷的气候条件，结合玉麦项目建筑材料运送距离长，运输难度大，通过前期调研及多轮方案对比。该项目民居建筑和陈列馆均采用冷弯薄壁型装配式钢结构，乡政府、村委会，小学、幼儿园，卫生院，公厕均采用装配式钢框架结构。

2．规划理念

（1）根据该项目的远期发展和需求进行设计，同时强调建筑、环境、景观及人的行为模式等诸多因素，合理组织，满足功能使用，注重土地开发的经济性，内外部交通、环境景观的协调性。

（2）在总体建设布局上，充分考虑了基地与周边环境及道路的关系，以及建筑在总体上的视觉效果，同时结合营造良好的总体空间与功能布局，营造出居住环境安静祥和的氛围；住宅建筑结合玉麦的气候特征布局，使得有良好的景观和日照。

（3）活动空间分级组织，场地中心公共区域：白塔、寺庙、篮球场、政府办公楼、民宿广场等公共空间开敞大气，给居民提供一个集会、娱乐健身等多种功能的活动场地。住宅组团围合成半私密半开敞的休闲活动空间，便于邻里交往，增进感情。注重公共活动空间的环境设计，处理好建筑、道路、广场、绿地和小品之间及其与人的活动之间的相互关系，丰富与美化环境。

（4）合理组织人流、车流和车辆停放，创造安全、安静、方便的居住环境；能使居民安居乐业，便于进行生活生产活动。

（5）满足用地规划和消防要求。

3．技术特色

该项目为边境小康村的建设，西藏自治区各级政府高度重视，深入贯彻科学发展观，坚持区域统筹发展，统筹安排村庄功能布局、公共服务设施、基础设施；统筹产业发展、统筹社会管理机制，促进区域全面协调可持续发展。充分尊重农牧民群众意愿和少数民族习俗，着力改善村民生产生活条件，切实让广大村民群众享受到小康乡建设的成果，是项目设计的重点，玉麦乡位于中印边境的藏南地区，从隆子县县城到玉麦乡驻地有6~9个小时的车程，只有一条10m宽的在建的边境公路通往玉麦乡驻地，沿途塌方落石情况较多，道路艰险，路况复杂，运输距离长，机械设备进出难度较大，大型运输货车无法进入，运输的难度及距离是施工的难度。民居建筑和陈列馆的楼盖及坡屋盖结构均采用冷弯薄壁型钢骨架组合楼板及屋面板。乡政府、村委会，小学、幼儿园，卫生院楼盖均采用压型钢承板上浇钢筋混凝土楼板，坡屋面均采用冷弯薄壁型钢骨架组合屋面板。可以有效地解决楼面减噪降噪的问题及玉麦多雨多雪对屋面荷载的要求。

冷弯薄壁型装配式钢结构和装配式钢框架结构自重轻，安装方便，施工工期较短。墙体、楼面、梁、屋架等大部分构件在工厂预制，每一个结构杆件都做到工业产品级精度，房屋安装万无一失，控制精度为毫米级。建筑工人从手艺人向装配技工转变，施工质量有保障。90%以上材料可回收、循环利用，消除建筑物终老时产生的大量建筑垃圾，实现"四节约""一环保"，建造过程绿色环保，可以有效地保护玉麦脆弱的高原生态环境。

4．结语

在玉麦项目施工的实际应用中得到进一步证实，装配式钢结构体系对现场作业要求低，工种较少，绝大部分作业在工厂中进行，机械化程度高，对工人的数量要求较低，通过前期的施工技术培训，工人的专业施工技术水平得到提升，施工周期短，可以有效地解决由于原材料紧缺及自然条件影响而造成无法大范围、快速建造的矛盾。对当地环境的保护，快速改善边境高寒地区生活居住条件，推动边境扶贫攻坚具有实质上的意义。

项目充分体现当地建筑文脉的延续性，根据现状建筑和传统藏式建筑特点，结合玉麦常年降雨量大，自然资源丰富等特点，建设特色鲜明、空间组织合理、功能齐全、尺度适宜的居住空间，形成独特的玉麦藏式建筑，让建筑可以适应当地的传统与发展，展现地域特色，风土人情。

Rural Construction in Ethnic Minority Areas under the Background of Rural Revitalization Strategy
——the Construction of Yumai Xiaokang Village, Yumai Township, Longzi County, the City of Shannan

Implementing the strategy of rural revitalization and coordinated regional development is a major task of building a modern economic system proposed by the 19th National Congress of the Communist Party of China. Adhering to the overall national security concepts, vigorously strengthening national defense construction, and maintaining border security are the major basic strategies for upholding and developing socialism with Chinese characteristics in the new era. Accelerating the construction of well-off townships in the border areas of Tibet, speeding up the poverty alleviation process, and realizing the grand goal of building a well-off society in an all-round way for villagers and the whole country is of great significance to the realization of the "two centenary goals" with Tibet, and its economic and social benefits and political benefits are significant. As our country has entered the decisive period of building a moderately prosperous society in an all-round way, the "Belt and Road" endeavors have been accelerated, regional coordinated development has been deepened, poverty alleviation has been promoted in full swing, the all-round national support for border areas has continued to increase, and the development of our relations with neighboring countries has entered into a new stage. There have been great chances to further promote the action of prospering national borders and people.

1. The Project Overview

Yumai Township is located in the southern foothill of Rila Mountain, Zari Mountain and Bosala Mountain, the northern branch of the Himalayas, and the southern part of the China-India border in the northern foothill of the Himalayas. The average altitude of Yumai Village is 3, 560 meters. The warm and humid Indian Ocean flow runs northward along the Yumai River Valley and is blocked by Rila Mountain with an altitude of 5, 000 meters, resulting in the climatic characteristics of Yumai being snowy in Winter and Rainy in Spring, Summer and Autumn. During the period from November to May, mountains are fully covered by heavy snow. The total land area of the project is 29. 40 hectares, the total construction area is 17,273. 25 square meters, and the total investment of the project is about 110 million yuan. After the project is completed, it will benefit 56 households with 192 local residents. The first-phase construction project includes 56 residences, a multi-storey township government and village committee, a multi-storey elementary school, a kindergarten, one multi-storey health center, a multi-storey exhibition hall, and two public toilets. The total construction area is 11,612.79m². According to the overall requirements of Tibet Autonomous Region for the construction of this project, and taking into account the rainy and cold climate conditions in Yumai Township, the long transportation distance and difficult transportation of construction materials for the Yumai project, preliminary investigations and multiple rounds of plan comparisons have been carried out. The residential buildings and exhibition halls of this project all adopt cold-formed thin-walled prefabricated steel structures, and the township government, village committees, elementary schools, kindergartens, health centers, and public toilets all adopt prefabricated steel frame structures.

2. Planning concept

(1) Westart to design to the long-term development and needs of the project, while emphasizing many factors such as architecture, environment, landscape and human behavior patterns, rational organization, satisfying functional use, focusing on the economy of land development, internal and external transportation, and environment coordination of the landscape.

(2) In the overall construction layout, the relationship among the base and the surrounding environment and roads, as well as the overall visual effect of the building, is fully considered. At the same time, the overall space and functional layout are combined organically to create a quiet and peaceful living environment; The residential buildings are arranged in combination with the climatic characteristics of Yumai so that there will be nice views and sunlight.

(3) Hierarchical organization of activity space, public areas at the center of the venue: a white pagoda, a temple, a basketball court, a government office building, a homestay inn, a square, and other public spaces are open and atmospheric, providing residents with a venue for gatherings, entertainment, fitness, and other functions. The residential groups are enclosed into semi-private leisure activity spaces, which are convenient for neighbors to communicate and enhance feelings. Pay attention to the environmental design of public activity spaces, handle the interrelationship among buildings, roads, squares, green spaces and miniature gardens and human activities so as to enrich and beautify the environment.

(4) Reasonably organize the flow of people, vehicle traffic and parking to create a safe, quiet and convenient living environment; enable residents to live and work in peace and contentment, and facilitate life and production activities.

(5) Meet the requirements from land planning and the fire protection requiement.

3. Technical Features

This project is the construction of a well-off village on the border. Governments at all levels in Tibet Autonomous Region attach great importance to it, thoroughly implement the scientific concept of development, adhere to overall regional development, and make overall arrangements for the functional layout of villages, public service facilities and infrastructure; overall industrial development, and overall social management mechanisms, as well as promote comprehensive, coordinated and sustainable development in the region. While fully respecting the wishes of farmers and herdsmen and the customs of ethnic minorities, the government has been focusing on improving the production and living conditions of villagers, and effectively allowing the majority of villagers to enjoy the results of the construction of a well-off township, which are the key points of the project design. Yumai Township is located in the southern Tibet area on the Sino-Indian border. It takes 6-9 hours to drive from Longzi County to Yumai Township. There is only a 10-meter-wide border road under construction leading to Yumai Township. There are many landslides and rocks along the way. The transportation distance is long, and it is difficult for the mechanical equipment to enter and exit, and the large-scale transport trucks cannot enter. The difficulty and distance of transportation are the difficulty of construction.

The roof and the sloping roof structures of residential buildings and exhibition halls are made of cold-formed thin-walled steel frame and roof panels. The township government, the village committee, the elementary school, the kindergarten, and the health center all use reinforced concrete floors cast with steel plates, and the sloping roofs are applied with cold-formed thin-walled steel frame and roof panels. It can effectively solve the problem of noise reduction on the floor and the requirements for roof load due to rain and snow.

Cold-formed thin-walled prefabricated steel structures and prefabricated steel frame structures are light in weight, easy to install, with a short construction period. Most of the components such as walls, floors, beams, and roof trusses are prefabricated in the factory, and each structural member achieves industrial product-level precision. The construction workers themselves have changed from craftsmen to assembly technicians, hence ensuring construction quality.. More than 90% of the materials can be recovered and recycled, hence eliminating a large amount of construction waste generated when the building becomes aged, realizing "four savings" and "one environmental protection", and the construction process is green and environmentally friendly, which can effectively protect the plateau of fragile ecological environment of Yumai.

4. Conclusion

It has been further confirmed in the practical application of Yumai project construction that the prefabricated steel structure system has low requirements for on-site operations and fewer types of work. The training of construction technology in the early stage has improved the professional construction technology level of workers, and the construction period is short, which can effectively solve the contradiction that the construction cannot be carried out on a large scale and quickly due to the shortage of raw materials and the influence of natural conditions. The protection of the local environment, the rapid improvement of living conditions in the alpine areas of the border, and the promotion of poverty alleviation in the border area are of substantial significance.

This project fully reflects the continuity of the local architectural context. According to the existing buildings and the characteristics of traditional Tibetan buildings, and in line with the characteristics of annual rainfall and abundant natural resources in Yumai, the construction features distinctive, reasonable spatial organization, complete functions, and appropriate scales. The living space forms unique Tibetan-style buildings in Yumai so that the buildings can adapt themselves to the local tradition and development, and demonstrate the regional characteristics and customs.

山南市隆子县玉麦乡玉麦小康村建设项目

Construction Project of Yumai Well-off village,
Yumai Township, Longzi County, Shannan City

用地面积:
294000m²
建筑面积:
17273.25m²
容积率:
0.06
设计时间:
2014-2015
建造时间:
2015-2017

玉麦乡位于喜马拉雅山北支脉日啦山、扎日山、博沙拉山南麓，喜马拉雅山主脉北麓的中印边境藏南地区。玉麦乡村驻地平均海拔 3560 m，印度洋暖湿气流沿玉麦河谷北上，遇到海拔 5000 m 的日啦山阻挡，造成玉麦冬季多雪、春夏秋三季多雨的气候特点。

项目总用地面积 29.40 公顷，总建筑面积 17273.25m²，项目总投资约 1.1 亿元。其中一期建设工程包含 56 栋民居，1 栋多层乡政府、村委会，1 栋多层小学、幼儿园，1 栋多层卫生院，1 栋多层陈列馆及两栋公厕，一期建设总建筑面积为 11612.79 m²。

（下图：山南市隆子县玉麦乡玉麦小康村建设项目总图）

Site Area:
294,000m²
CFA:
17,273.25m²
FAR:
0.06
Design Time:
2014-2015
Construction Time:
2015-2017

Yumai Township is located at the southern foot of Rila Mountain, Zari Mountain and Bosala Mountain, the northern branch of the Himalayas, and the southern part of the China-India border at the northern foot of the Himalayas. The average altitude of Yumai Village is 3,560m. The warm and humid Indian Ocean flow running northward along the Yumai River Valley is blocked by Rila Mountain with an altitude of 5,000m, resulting in the climatic characteristics of Yumai being snowy in winter and rainy in spring, summer and autumn. The total land area of the project is 29.40hm², the total construction area is 17,273.25m², and the total investment of the project is about 110 million yuan. The first-phase construction project includes 56 residences, a multi-storey township government and a village committee, a multi-storey elementary school, a kindergarten, a multi-storey health center, a multi-storey exhibition hall, and two public toilets. The total construction area is 11,612.79m².

(Below: A master plan of the construction project of Yumai Xiaokang Village, Yumai Township, Longzi County, Shannan)

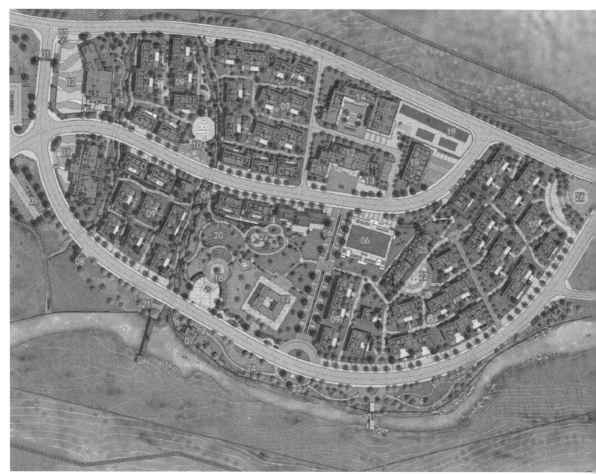

设计愿景：通过边境小康乡建设，改善乡村风貌，修复生态环境，并融合当地民族特色，营造具有玉麦地域特色的自然及人文景观，打造山水相依、民族文化丰富，满足不同人群游赏、活动需要的优山美村。完善景观配套设施，使其成为爱国主义教育基地。

Design visions: Through the construction of a well-off town c the border, efforts are made to improve the rural style, resto the ecological environment, and integrate the local ethn characteristics to create a natural and cultural landscape with t characteristics of Yumai region, create a marvelous village marke by mountain and river-dependence, rich ethnic culture, a satisfying different groups of people for sightseeing and activitie Improve the landscape supporting facilities and make it a base f patriotic education.

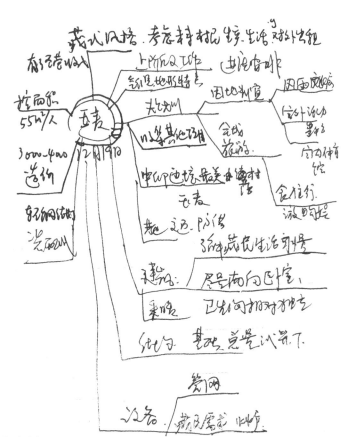

根据西藏自治区的总体要求，考虑到玉麦乡多雨、寒冷的气候条件，结合玉麦项目建筑材料运送距离长，运输难度大，通过前期调研及多轮方案对比。该项目民居建筑和陈列馆均采用冷弯薄壁型装配式钢结构，乡政府、村委会，小学、幼儿园，卫生院，公厕均采用装配式钢框架结构。

According to the general requirements of Tibet Autonomous Region, considering the rainy and cold climate conditions in Yumai Township, combined with the long transportation distance and difficult transportation of building materials in Yumai project, through preliminary investigation and comparison of multiple rounds of schemes, the residential buildings and exhibition halls of the project adopt cold-formed thin-walled fabricated steel structure, and the township government, the village committee, the elementary school, the kindergarten, the health center, and the public toilet adopt fabricated steel frame structure.

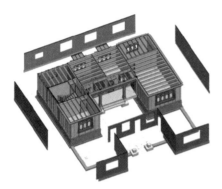

基于 BIM 技术对项目进行配饰式深化拆分指导。对搭建好的整体 BIM 模型进行拆分，拆分工作主要针对外墙和楼板，根据窗口、门洞的位置进行拆分设计。并对连接处进行深化设计，保证拆分构件的预埋设计准确无误。基于 BIM 模型对已完成的主体轻钢结构进行编号及组合，对材料统计及现场拼装起到一定的作用。

建筑采用装配式钢结构体系，通过对建筑细部的精心优化设计，使其不但具有保温防潮效果好，抗震性能优的特点，而且符合藏族传统建筑的特征和审美价值。

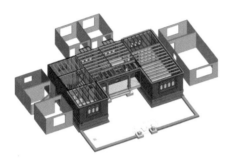

Based on BIM technology, the project is guided by accessory-style deepening and splitting. Split the built overall BIM model. The splitting work is mainly for the exterior walls and floor slabs, and the splitting design is carried out according to the positions of window and door openings. Deepen the design of the connection to ensure the accuracy of the embedded design of the split components. Based on the BIM model, the completed main light steel structure is numbered and combined together, which plays a considerable role in material statistics and on-site assembly.

The building adopts the prefabricated steel structure system. Through devoted optimization in the building details, it not only has the characteristics of good thermal insulation and moisture-proof effect, excellent seismic performance, but also is in line with the characteristics and aesthetic value of traditional Tibetan architecture.

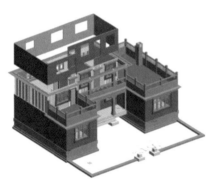

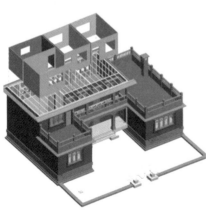

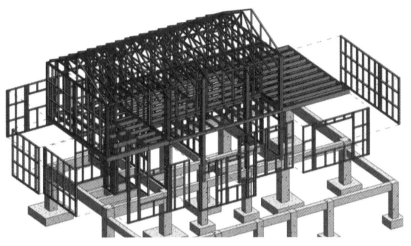

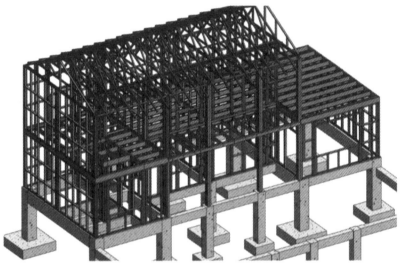

轴线与斑块
Axis and plaque

浏阳护理小镇
Liuyang Nursing Town

除拟规划的书院区建设区域外，其余建设区域均为山地，地形起伏较大，如何在最大限度尊重自然的前提下结合地形地貌，充分发掘场地潜力，延续地形地势特色实现"传统民居村落型学校"的设计理念是我们的首要问题。采用柔畅飘逸的轴线，一气呵成地将各个建筑串联在一起，总体建筑布局自东向西呈现出动感姿态，蕴含"治愈绿谷"的深刻含义。生机盎然的绿色以及山形的弧度赋予其柔美的一面；同时，群山对谷地的环抱又使人联想到对生命的保护。这两种形象正像是对医院特性的隐喻：以人为本，呵护生命，治愈之谷。

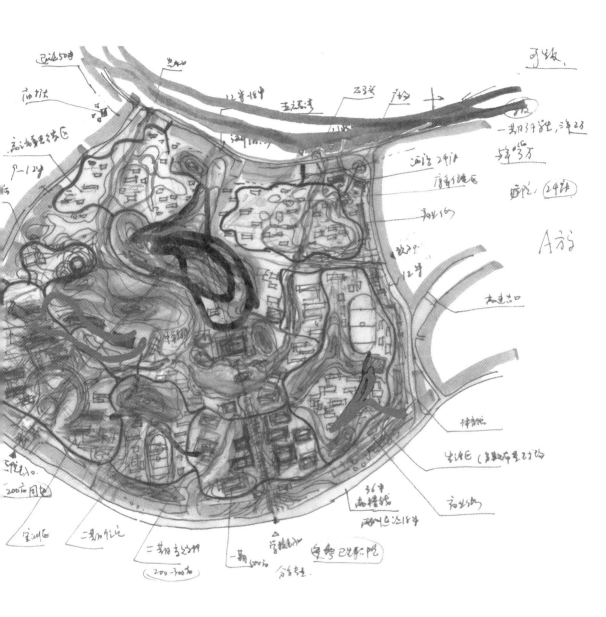

小满

夜莺啼绿柳，皓月醒长空。

——宋·欧阳修《五绝·小满》

湖南护理学校浏阳校区及学校附属医院规划设计研究

随着社会经济的发展，一些社会问题诸如常见疾病谱扩大，生活方式不健康，人口老龄化等问题日益严重，健康成人们最重视的目标，医疗工作体现出社会重要性。护理工作是医疗工作的重要组成部分，在当今竞争日益激烈的医疗市场中，护理质量的好坏直接影响了医疗水平的高低，常言道"三分治疗，七分护理"，这句话虽然并不十分准确，但却反映出护理工作的重要作用和地位。当今护理学随着科学的发展，已经成为一门综合性多学科的应用科学。

湖南护理学校浏阳校区将建设成为多维度多层次的现代化护理高等专科学校，近期目标为：成了一所开设临床学院、口腔学院、护理学院、药学院、医技学院等十二个学院以上的，以医卫专业为主的、综合性的、非营利性的高职学院。远期 5~10 年后扩展目标将新增七个以上医护专业，最终到达一所以医卫专业为主的、综合性的、非营利性的本科院校。

1. 项目概况

项目位于浏阳市官桥镇八角亭村，距离浏阳市西南侧约 30km 处，平汝高速以西，南横线以北，浏阳市是湖南省辖县级市，由长沙市代管，因县城位于浏水之北（阳面）而得名。地块内地势为丘陵地带，地理环境优越，周边交通便利，环境优美，临近浏阳河，发展前景远大。

该次项目的设计内容为湖南护理学校浏阳校区方案设计，设计范围为总图、建筑、结构、给水排水、电气、暖通等。总建筑面积约为：533949.39m²，主要工程内容有公共教学楼、实验楼、礼堂、图书馆、创业楼、行政楼、风雨操场、体育馆、食堂、学生宿舍、学术交流中心、实训中心，以及大门等辅助用房。

2. 规划理念

该次校园用地面积约 440666.7m²（661 亩），具体地形位于丘陵地带，形状十分不规则，通过严谨现状分析，规划将学校主入口放在南侧，靠近主要道路，次入口分别设置在靠近学校一期生活区域的西侧和二期生活区的东侧，学校主入口处设置开敞的中心广场，是师生人流的集散空间，也是学校治学形象的展示空间，结合景观设计，把广场打造为学校的室外精神课堂，将知识学习赋予神圣的仪式感，并由此形成一条南北向的中心轴线（共享区）。

学校的行政楼和创业楼设置在入口两侧位置，地块的东西两侧分别设置了一期和二期独立组团的学生生活和运动区域，便于学校分期开发和独立使用。学校的主要教学功能用房均南北向布置，为教学提供良好的条件。实训楼通过连廊连接，为师生在不利气候条件下的通行提供便利条件。

在提倡素质教育的当下，学校教学的目的已经不仅仅在于传道授业解惑，更重要的是激发起学生对于知识学习的兴趣，使其可以乐学其中，提升整体素质。因此传统的教育中相对独立的单功能的教学空间已经不能满足当下教育中越来越强调的学科融合，现代教育更注重综合性和启发性，也更依赖生态性、体验性的复合型空间。

基于这样的思考，在该项目中设计提出"乐学大院"的空间概念模式。所谓"乐学大院"，简单来说，未来的校园既是"学园"，也是"乐园"，即帮助学生可以更好地从课堂获取知识，同时通过灵活多样的功能设置以及多维度的空间设计，丰富学生课堂内外的活动，打造具有积极意义的生活空间，鼓励学生乐学向上，奋勇拼搏，通过互动交流赋予校园更有场所含义的学习型综合体概念——快乐起航，青春绽放。

该项目在建筑设计上借鉴欧美哈佛大学、剑桥大学的成功理念，吸纳岳麓书院、湘雅医院、北京协和医院等建筑精髓，以"洋为中用，古为今用"为设计理念，力争建设成为突出中华文化特点，兼容中外文化元素，突显湖南本土特色的现代化校园。

3. 建筑设计

（1）文脉传承的植入

每个建筑在当地的植入都要与当地文化传统相结合，追寻当地文化脉络。沿着场地的中轴线，我们在室外环境中设计了展现学校历史文化的现代大学的特色景观空间，并且还可以作为课余时间休息、阅读的活动区间，丰富学生的课外活动。

（2）建筑朝向的处理

整个建筑群布局采用建筑堪舆学的五行相环、和谐共生的布局，体现出学校和谐共荣的特点。建筑整体正南方向偏东约 15°，整体走向呼应周边道路及周边整体建筑走向，使块紧密联系，与周边界面形成整体。

（3）交流空间

场地各区域由建筑与连廊共同限定出不同属性的公共空间，为多样性的校园生活提供与之适应的场所。我们力图在这里塑造一个核心，这一空间意义上的核心促进了学生们的各种交流，并加强了学校作为一个大家庭的归宿感。提供给逗留、集会和相遇创造大量机会的空间。形成街道、庭院、广场等拟城市的空间类型，促进了校园的社交行为。

4. 结语

湖南护理学校新校区项目充分利用该区域优越的交通环境和建设方的专业底蕴，全方位展现清华大学、岳麓书院、湘雅医院、北京协和医院等建筑精髓、建筑风情、生态环境和生活方式，引入"乐学大院"的空间概念模式，将现代化教学和旅游、休闲、文化、生活等有机结合，符合湖南城市建设和产业发展要求，对促进浏阳及长沙的发展产生极大推动作用。

Planning and Design of Liuyang Campus of Hu'nan Nursing School and Its Affiliated Hospital

With the development of social economy, some social problems such as the expansion of common disease spectrum, unhealthy lifestyle, aging population, etc., are becoming more and more serious. Health has become the most important goal of people, and medical work reflects the importance of society. Nursing is an important part of medical work. In today's increasingly competitive medical market, the quality of nursing directly affects the level of medical care. As a saying goes that 30% of efforts on treatment and 70% on nursing, not accurate as it might be, it reflects the important role and status of nursing. With the development of science, today's nursing has become a comprehensive multi-disciplinary applied science.

Liuyang Campus of Hu'nan Nursing School will be built into a multi-dimensional and multi-level modern nursing college. The short-term goal is to open more than a dozen clinical colleges, oral colleges, nursing colleges, pharmacy colleges, and medical technology colleges, and to be a professional, comprehensive, non-profit vocational college. In five to ten years, more than seven medical and nursing majors will be added. Finally, it will become a comprehensive, non-profitable undergraduate college focusing on medical and health majors.

1. The Project Overview

The project is located in Bajiaoting Village, Guanqiao Town, the City of Liuyang , more than 30 kilometers away from the southwest of Liuyang, west of Pingru Expressway and north of Nanheng Line. Liuyang, as a county-level city under the jurisdiction of Hu'nan Province, is managed by the City of Changsha. It is named after the north of Liushui (the sunny side). The terrain in the plot is a hilly area, the geographical environment is superior, the surrounding transportation is convenient, the environment is beautiful. It is close to the Liuyang River.

The content of this project is the schematic design of Liuyang Campus of Hu'nan Nursing School, including general plan, architecture, structure, water supply and drainage, electrical, HVAC, etc. The total construction area is about 533949. 39m^2, the main project includes a public teaching building, a laboratory building, an auditorium, a library, an entrepreneurial building, an administrative building, a wind and rain playground, a gymnasium, a canteen, a student dormitory, an academic exchange center, a training center , and auxiliary rooms such as the gate room.

2. The Planning Concept

The campus covers an area of over 44 hectares. The specific terrain is located in a hilly area, with a very irregular shape. Through rigorous analysis of the current situation, the main entrance of the school is planned on the southern side, close to the main road, and the secondary entrances are located on the western side of the living area of the first phase of the school and on the eastern side of the second-phase living area. An open central square is set at the main entrance of the school, which is a major socializing space for teachers and students, and a display space for the school's academic image. In line with the landscape design, the square is built as the school's outdoor spiritual classroom where knowledge learning is endowed with a sense of sacred rites. Accordingly, a central axis in the north-south direction was formed (a shared area).

The administrative building and entrepreneurial building of the school are located on both sides of the entrance. The eastern and western sides of the plot are respectively set up for the first and second phases of independent student living and sports areas, which are convenient for the school to develop and use independently. The main teaching classrooms of the school are arranged in a north-south direction, providing good conditions for teaching. The training buildings are connected by corridors to provide convenience for teachers and students under unfavorable weather conditions.

At present, when quality education is advocated, the purpose of school teaching is not only to teach and solve doubts, but also to stimulate student interest in knowledge learning so that they can enjoy learning and improve their overall quality. Therefore, the relatively independent single-function teaching space in traditional education can no longer satisfy the integration of disciplines which is increasingly emphasized in current education. Modern education pays more attention to comprehensiveness and inspiration, and also relies more on ecological and experiential compound spaces.

Based on this kind of thinking, in this project, we design and propose the spatial concept model of "Happy Academy". The so-called "Happy Academy", in simple terms, means that in the future the campus will be both a "school" and a "paradise", i.e. to help students better acquire knowledge from classrooms and at the same time, through flexible and diverse functional settings and multi-dimensional spaces design, to enrich student activities inside and outside classrooms, to create a positive living space, to encourage students to be happy to learn, to work hard, and to give the campus a more place-based learning complex concept through interactive exchanges, i.e. happy sailing, youth blooming.

In architectural design, this project draws on the successful concepts of Harvard University in the United States and Cambridge University in Britain, and absorbs the architectural essence of Yuelu Academy, Xiangya Hospital, Peking Union Medical College Hospital, etc. Following the design concept of "learning from foreign countries and the past" , efforts are made to bring out Chinese cultural characteristics, integrate Chinese and foreign cultural elements, and build a modern campus that highlights the local characteristics of Hu'nan Province.

Architectural design

Implantation of cultural heritage

The local implantation of each building must be combined with local cultural traditions and pursue the local cultural context. Along the central axis of the site, we designed the characteristic landscape space of a modern university in the outdoor environment to showcase the history and culture of the school, and it can also be used as an activity area for rest and reading in spare time to enrich extracurricular activities of students.

2) Handling of building orientation

From the perspective of architectural Fengshui, the layout of the entire building group adopts five elements interlocking and harmonious coexistence layout of architectural Fengshui, which reflects the characteristics of the school harmony and co-prosperity. The overall southward direction of the building is diverted about 15° east, and the overall direction agrees with the surrounding roads and the surrounding overall building direction, so that the blocks are closely connected, forming an entity with the surrounding interface.

3) Communication space

Each area of the site is jointly defined by buildings and corridors to define public spaces with different attributes, providing suitable places for diverse campus life. We strive to build a core here, a spatial core that facilitates various exchanges among students and reinforces the sense of belonging of the school as a family. We provide spaces that offer a multitude of opportunities for staying, gathering and encountering. The space types such as streets, courtyards, and squares are formed, hence promoting social behaviors of the campus.

4. Conclusion

The new campus project of Hu'nan Nursing School makes full use of the superior traffic environment in the area and the professional background of the construction party, and fully displays the architectural essence, architectural style, ecological environment, and lifestyle of Tsinghua University, Yuelu Academy, Xiangya Hospital, Peking Union Medical College Hospital, etc. The introduction of the spatial concept model of "Happy Academy", which organically combines modern teaching with tourism, leisure, culture, and life, meets the requirements of urban construction and industrial development in Hunan Province, and will greatly promote the development of Liuyang and Changsha.

浏阳护理小镇

Liuyang Nursing Town

用地面积：
1266666.7m²
建筑面积：
1071760m²
容 积 率：
0.85
设计时间：
2016- 至今
建造时间：
2020- 至今

项目总用地约为 1266666.7m²（1900 亩），场地为原始山地丘陵地貌，地形较为复杂，地势整体呈北高南低。该项目以大健康产业为基础，以医疗培训为主导，打造一个以医疗培训、疗养康复、养生社区、休闲体验等精品项目为支撑的湖湘医疗康养度假目的地。其中护理学校总建筑面积 561714.8m²、医院总建筑面积 137419.3m²、康养总建筑面积 87550m²、商住总建筑 285075.8m²。结合康养领域融合发展构筑复合型医疗康养护理小镇。

（下图：浏阳护理小镇总图）

Site Area：
1,266,666.7m²
GFA：
1,071,760m²
FAR：
0.85
Design Time：
2016- To date
Construction Time：
2020- To date

The total land area for the project is approximately 1,266,666.7m²(1,900mu), situated on hilly terrain with complex topography and a generally north-to-south slope. The project is centered around the large health industry, with medical training as the main focus, creating a premium destination for medical training, recuperation and rehabilitation, health preservation communities, and leisure experiences in the Hunan-Xiangtan Medical Health and Recuperation Resort. The nursing school has a total construction area of 561,714.8m², the hospital has a total construction area of 137,419.3m², the healthcare and recuperation facilities have a total construction area of 87,550m², and the commercial and residential spaces have a total construction area of 285,075.8m². By integrating development in the healthcare and recuperation fields, a composite medical health and nursing small town is being built.

(Below: Overall map of Liuyang Nursing Small Town)

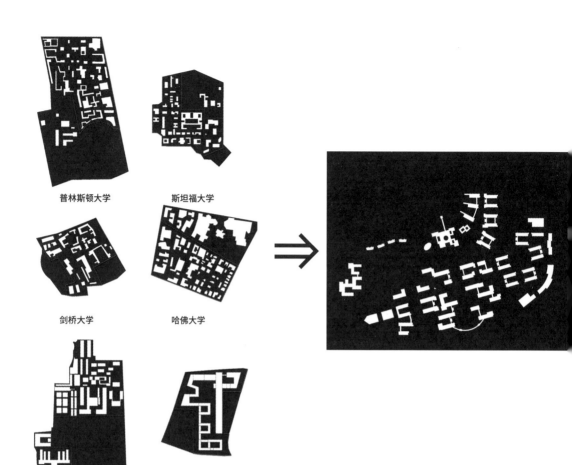

普林斯顿大学　　　斯坦福大学

剑桥大学　　　哈佛大学

岳麓书院　　　株洲中等职业学校

建筑基地关系的设定对于形成宜人尺度关系的空间环境具有实质性影响，规划方案的基地关系以研究国际著名学校及长沙相关卫生护理学校为参照，规划在避让高压线，保留农田的基础上结合自然生态环境，最终形成生长、连续、舒适自然的学校基底关系。

The setting of the relationship among the building bases has a substantial impact on the spatial environment that forms a pleasant scale relationship. The base relationship of the scheme is based on the study of internationally renowned schools and related health care schools in Changsha. Under the principle of avoiding the high-voltage electric wire and reserving the farming land, the school-base relationship which is growing, continuous, and comfortable was finally formed.

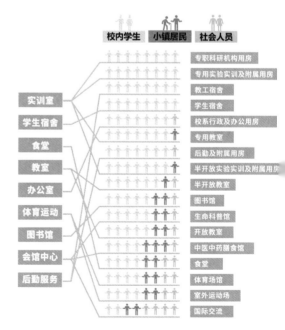

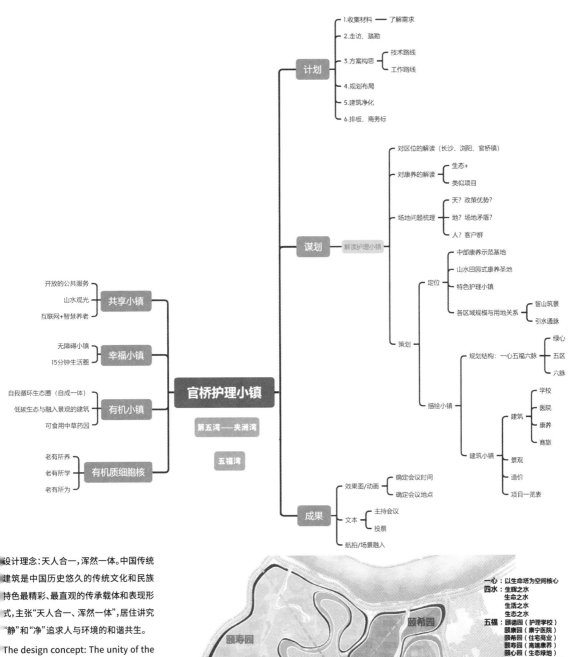

计划
- 1.收集材料 —— 了解需求
- 2.走访、踏勘
- 3.方案构思 —— 技术路线 / 工作路线
- 4.规划布局
- 5.建筑净化
- 6.排板、商务标

谋划 —— 解读护理小镇
- 对区位的解读（长沙、浏阳、官桥镇）
- 对康养的解读 —— 生态+ / 类似项目
- 场地问题梳理 —— 天? 政策优势? / 地? 场地矛盾? / 人? 客户群
- 策划
  - 定位
    - 中部康养示范基地
    - 山水田园式康养圣地
    - 特色护理小镇
    - 各区域规模与用地关系 —— 留山筑景 / 引水通脉
  - 描绘小镇
    - 规划结构：一心五福六脉 —— 绿心 / 五区 / 六脉
    - 建筑小镇
      - 建筑 —— 学校 / 医院 / 康养 / 商旅
      - 景观
      - 造价
      - 项目一览表

成果
- 效果图/动画 —— 确定会议时间 / 确定会议地点
- 文本 —— 主持会议 / 投票
- 航拍/场景融入

官桥护理小镇
第五湾——夹洲湾
五福湾

共享小镇
- 开放的公共服务
- 山水观光
- 互联网+智慧养老

幸福小镇
- 无障碍小镇
- 15分钟生活圈

有机小镇
- 自我循环生态圈（自成一体）
- 低碳生态与融入景观的建筑
- 可食用中草药园

有机质细胞核
- 老有所养
- 老有所学
- 老有所为

设计理念：天人合一，浑然一体。中国传统建筑是中国历史悠久的传统文化和民族特色最精彩、最直观的传承载体和表现形式，主张"天人合一、浑然一体"，居住讲究"静"和"净"追求人与环境的和谐共生。

The design concept: The unity of the nature and man. traditional Chinese architecture is the most wonderful and intuitive inheritance carrier and expression form of traditional Chinese culture and national characteristics with a long history. It advocates the "Unity of the nature and man", pays attention to "quiet" and "clean" way of living, and pursues harmonious coexistence between man and the environment.

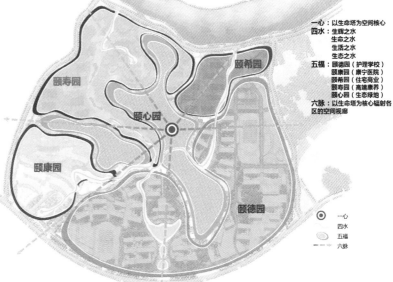

一心：以生命塔为空间核心
四水：
- 生辉之水
- 生命之水
- 生活之水
- 生态之水
五福：
- 颐德园（护理学校）
- 颐康园（康宁医院）
- 颐希园（住宅商业）
- 颐寿园（高端康养）
- 颐心园（生态绿地）
六脉：以生命塔为核心辐射各区的空间视廊

一心
四水
五福
六脉

通过医疗街将综合医院和三大疗养中心形式交通上的串联,实现医院内部步行交通的完整性;并在重要节点处设置休闲留驻空间,满足人行交通的不同需求。

医院内部通过不同等级的人行交通实现步行系统的完整性,通过入口广场实现外部车行交通与医院内部的连接,通过游憩型道路实现医院景观环境的共享。

将水系引入院内,水脉贯通,内外水系成为一体,使绿化成为医院景观的中心,引入周边两侧山体景观,构建完整景观体系。

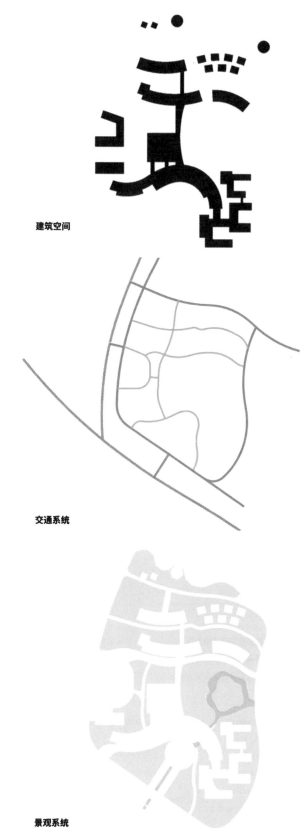

建筑空间

交通系统

景观系统

The comprehensive hospital and three major health centers are connected by the Medical Street, hence ensuring the integrity of the pedestrian traffic inside the hospital. Leisure spaces are set up at key locations to meet varied needs of pedestrian traffic.

The interior of the hospital is based on the integrity of different levels of the pedestrian traffic system. The external plaza is connected with the interior of the hospital through the entrance plaza, and the hospital landscape environment is shared through the recreational road.

The water-system was introduced into the hospital, and the internal and external water systems were integrated, making the green scenery the center of the hospital landscape. Together with mountain scenes on both sides, a scenery landscape has been completed.

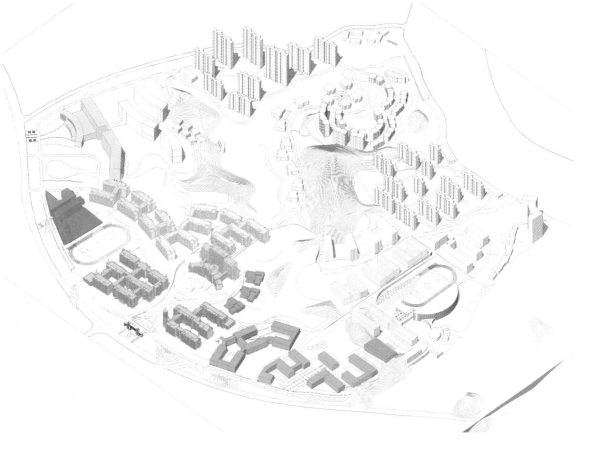

打造人性化的交往场所保护生态、自然的山水校园蕴含奉献、护理精神的护理学校。叠石而落，山间庭院。学校设计结合空间的多维文化感应，营造轻松和谐的学习氛围，基于人性化的构思及依地势叠石而落的印象，将美感延展为一个运动的过程。因势生长，自然建筑。结合基地地形进行底层架空、平台跌落等创新型设计，在塑造学校浓厚学习氛围同时，更好地与自然地形相结合。

Create a humanized place for communication and protect the ecology and nature. The campus contains a nursing school with dedication and a nursing spirit. Falling on top of the rocks, the courtyard is amidst mountains. The school design combines the multi-dimensional cultural sense of the space to create a relaxed and harmonious learning atmosphere. Based on the humanized conception and the impression of falling rocks on top of the terrain, the beauty is extended into a process of movement. Growing in response to the trend, natural architecture. Innovative designs such as overhead and platform drops are carried out in combination with the topography of the base. While shaping the strong learning atmosphere of the school, it is better integrated with natural topography.

规整与自由

Order and freedom

张家界福苑皇家国际度假酒店
Zhangjiajie Fuyuan Royal International Resort Hotel

季节性是旅游活动最显著的特征之一，这种季节性波动往往会造成酒店不必要的浪费。为应对旅游季节性不平衡，我们从淡季和旺季两种情景出发，根据场地特色和酒店自身运营特点的综合考虑，总平面采用"一轴两翼"对称式的整体布局方式。主轴垂直于高云路方向，由前入口广场、酒店大堂、水幕景观，中间庭院串联形成酒店主要的公共交流空间。沿主轴两翼布置酒店各项服务功能。起初的设想是传统民宿与现代酒店为两翼，双方既可以同时开放也可以单独开放，体现着自由与规整的有机结合，一边是传统村落，承载着盛世乡愁，还原本土风貌；一边是现代酒店，展现着现代智慧，满足多样化需求。经过沟通与修改，我们与甲方达成一致，改成两边都是现代酒店，中间穿插着民宿，用中式传统景观格局、尽显经典雅致。加上自然式的水体与亭、台、植物的精致搭配，让整个酒店庭院散发出休闲优雅的独特品位。

芒种

荷风送香气，竹露滴清响。

——唐·孟浩然《夏日兰亭怀辛大》

张家界福苑皇家国际度假酒店

Zhangjiajie Fuyuan Royal International Resort Hotel

用地面积：
55752m²
建筑面积：
66174.55m²
容 积 率：
0.85
设计时间：
2010-2011
建造时间：
2012-2019

张家界福苑皇家国际度假酒店位于张家界武陵源区高云路西北侧，项目定位为五星级标准度假酒店。高云路是由张家界市区进入武陵源各风景区的必经之路。项目场地位于高云路的关键节点位置，是旅游休闲集散的黄金位置。

项目总用地面积为 55752m²，约合 83.42 亩，基地呈 L 形，东西向 255m，南北长 309m，基地由南向北为缓坡，高差约为 2m，靠西北向为山丘，山体场地相对高差 20~30m。

整体布局体现出尊重自然、利用自然与尊重地方文化同时再创新的设计原则。根据场地特色和酒店自身运营特点的综合考虑，总平面采用 一轴两翼对称式的整体布局方式。主轴垂直于高云路方向，由前入口广场、酒店大堂、水幕景观，中间庭院串联形成酒店主要的公共交流空间。沿主轴两翼布置酒店各项服务功能。

（下图：张家界福苑皇家国际度假酒店总图）

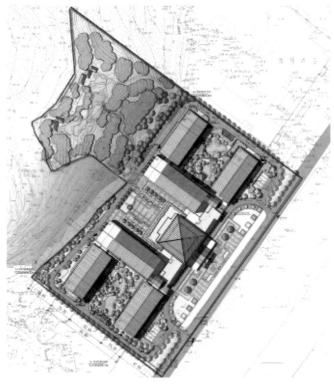

Site Area:
55,752m²
CFA:
66,174.55m²
FAR:
0.85
Design Time:
2010-2011
Construction Time:
2012-2019

The total land area of the project is 55,752m², about 83.42mu. The base is L-shaped, 255m east to west and 309m north to south. The base is a gentle slope from south to north, and the height difference is about 2m. At the northwest side are hills. The relative height difference of the mountain site is 20~30m. The overall layout reflects the design principle of respecting the nature, and innovation based on using the environment and respecting local culture. According to the site characteristics and the operation characteristics of the hotel, the general plane adopts an overall symmetrical layout of one axis and two wings. The main axis is perpendicular to Gaoyun Road. With the front entrance square, the hotel lobby, the water curtain landscape, and the middle courtyard contribute to the main public communication space of the hotel. Various service functions of the hotel are arranged along the main axis and two wings.

（Below: A master plan of Zhangjiajie Fuyuan Royal International Holiday Inn）

项目总建筑面积 66174.55m²，其中地上建筑面积 44936.36m²，地下建筑面积 21238.190m²。地面建筑层数 5 层（局部 6 层），地下 1 层，建筑高度 23.90m。

一层：在建筑临高云路一侧正中布置酒店大堂，大堂左翼前侧为会议中心，设有大小会议室 6 间，配套有茶水间及商务中心等；左翼后侧为商务商业区和员工宿舍，大堂右翼前侧为餐饮区，设有 830m² 中餐大厅、西餐厅；右翼后侧为健身娱乐区，设有 200m² 室内游泳池，足浴房、健身房、形体塑造中心、美容美发等。

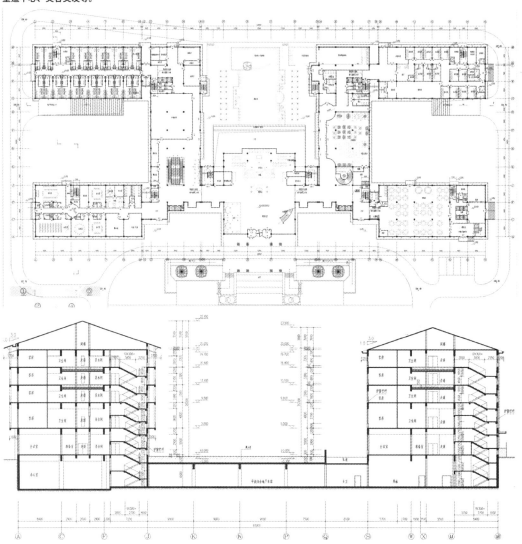

The total construction area of the project is 66,174.55 m², including 44,936.36 m² above the ground and 21,238.190m² underground. There are 5 floors above the ground (6 floors locally), 1 floor underground, and the building height is 23.90m.

The first floor: the hotel lobby is arranged in the middle of the side of the building facing Gaoyun road. The front of the left side of the lobby is the conference center, with 6 conference rooms, a tea room and a business center; At the back of the left wing is the business area and staff dormitory, and in front of the right wing of the lobby is the catering area, with a 830m² Chinese restaurant and a Western restaurant; at the back of the right wing is the fitness and entertainment area, with a 200m² indoor swimming pool, a foot massage room, a gym, a body-shaping center, a beauty salon, etc.

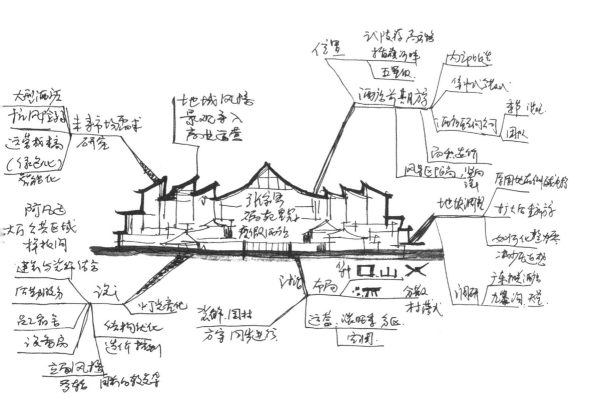

文绿相融

Green harmony

浔龙河生态示范点公共配套项目
Public Supporting Project of Xunlong River Ecological Demonstration Site Exhibition Hall

有一个地方，森林覆盖率超过70%，金井河、麻林河、浔龙河，三河穿梭，流淌在山林丘壑，……革命先烈杨开慧、国歌词作者田汉等名人生于斯、长于斯……优美生态、人文故事，宛如一幅厚重鲜活的江南水墨，沁人心脾。这就是浔龙河，位于长沙县果园镇。根据浔龙河的山水人文提炼出"龙、谷、田、歌、绿、园"六大规划概念元素，通过发散式创意将六大元素融入规划设计中，谱写"欢快动感、悠然写意、心灵超脱的田园生活三重曲"，形成了龙之谷、田之歌、绿之园三大功能分区。浔龙河的美，如惊鸿一瞥，来不及细细品味。在这里，你能感受到厚重的历史文化，享受闹市难见的宁静恬适。在这里，设计师是带着温度和感情下乡，原汁原味的乡愁得到保护。城中有村，村中有城；闹中求静，动静相宜，让村民诗意地栖居，让乡愁得以保留和延续。

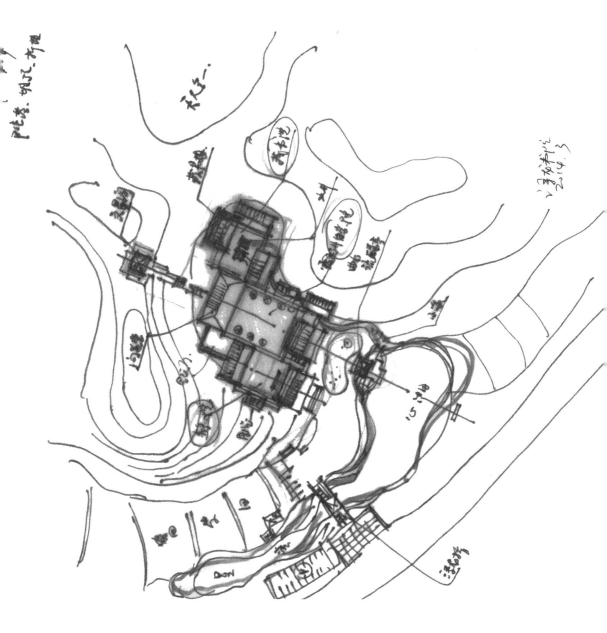

夏至

雨砌蝉花粘碧草，风檐萤火出苍苔。

——明·刘基《夏日杂兴·夏至阴生景渐催》

浔龙河生态示范点公共配套项目

Public supporting project of Xunlong River Ecological Demonstration Site

用地面积：

262603m²

建筑面积：

7173m²

容 积 率：

0.03

设计时间：

2013-2014

建造时间：

方案

浔龙河生态小镇项目位于长沙县腹地果园镇双河村。区位优越，气候宜人、地形秀美，浔龙河、金井河、麻林河、哑河多条水系交织环绕，多河汇聚成岛。狮子山气势恢宏，走马山悬崖陡壁，灵山秀水。地形地貌独特，整体呈现出"两多两少"的特点，即"山多、水多、田少、人少"。村内自然资源丰富，山清水秀，大片竹林、树林层岚叠嶂；水系尤其发达，浔龙河、金井河、麻林河三条河流交织环绕，与典型的江南丘陵地形地貌，互为映衬。

（下图：浔龙河生态示范点公共配套项目总图）

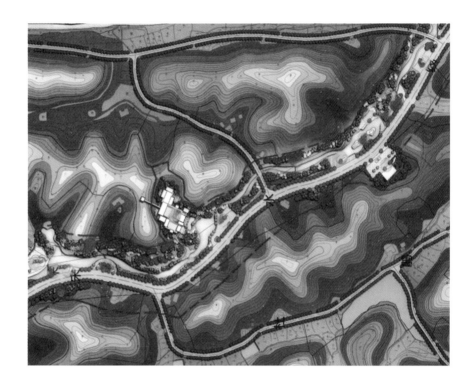

Site Area:

262,603m²

CFA:

7,173m²

FAR:

0.03

Design Time:

2013-2014

Construction Time:

Program

Xunlong River Ecological Town Project is located in Shuanghe Village, Guoyuan Town, the hinterland of Changsha County. The location is superior, the climate is pleasant, and the topography is beautiful. Xunlong River, Jinjing River, Malin River and Yahe River are surrounded by multiple water systems, with many rivers contributing to an island. Lion Mountain Rock is magnificent, with steep cliffs in Zoumashan Mountain and beautiful mountains and waters. The topography is unique, showing the characteristics of "two more's and two few's", i.e." more mountains, more water, few fields and few people". The village is rich in natural resources, beautiful mountains and limpid waters, large bamboo forests and massive forests. The water system is especially developed, with three intertwining rivers of Xunlong River, Jinjing River and Malin River, which complement one another with typical Jiangnan hilly topography.

（Below: A master plan of public supporting projects of Xunlong River Ecological Demonstration Site）

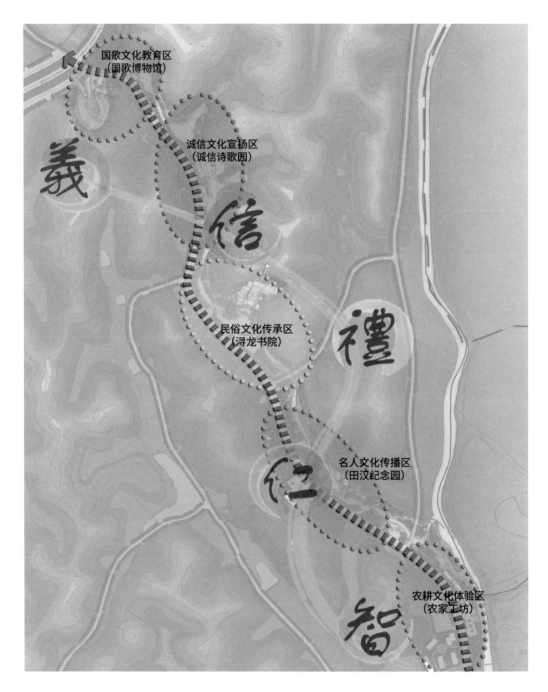

国歌文化教育区
（国歌博物馆）

诚信文化宣扬区
（诚信诗歌园）

民俗文化传承区
（浔龙书院）

名人文化传播区
（田汉纪念园）

农耕文化体验区
（农家工坊）

发展思路：以文化为魂、资源为根的理念，以两型社会、城乡统筹为指导思想，以生态、亲民、随和为意见，以景区化的理念打造景观风貌和环境，为村民创造一个良好的生活环境、发展环境，给游客描绘一个动听的故事、一幅漂亮的图画！

Development ideas: Taking culture as the soul and resources as the root, taking the two-type oriented society and the overall planning of urban and rural areas as the guiding ideology, taking ecology, being close to the people, and easy-going as opinions, to create a landscape and environment with the concept of scenic spots, and a nice living environment for villagers. Develop the environment, tell a beautiful story and draw a beautiful picture for tourists!

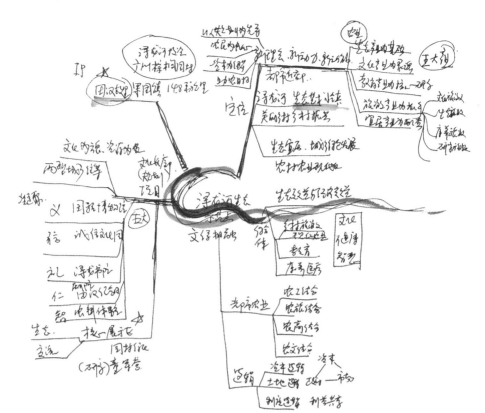

浔龙书院位于文化长廊以西，用地呈不规则形状，红线内东西向最宽约 173m，南北向长约 102m，地形起伏较大，总用地面积 1.00 公顷，整个地块依山傍水，视线、景观资源效果好。浔龙书院依托良好的天然地理优势，建筑设计风格以周家大院为代表的湖湘特色为主要元素，打造出一个集文化展览、艺术交流、学术讲座等功能为一体的地标性建筑。

Xunlong Academy is located to the west of the Cultural Corridor. The land is irregular. The red line is about 173m wide in the east-west direction and about 102m long in the north-south direction. The topography is undulating and the total land area is 1. 00hm². The entire plot is surrounded by mountains and waters. The effect of sights and landscape resources is good. Xunlong Academy relies on good natural geographical advantages, and its architectural design style is based on Huxiang characteristics represented by the Zhou Family Courtyard. It has created a landmark building integrating cultural exhibitions, art exchanges, academic lectures, and other functions.

国歌博物馆位于东八线与文化长廊交叉路口南向，地线高差较大，呈不规则三边形。用地总面积为 12683.22m²，东西最长约 130m，南北最长160m。国歌博物馆的设计体现了《义勇军进行曲》精神内涵："凝聚"建筑造型积聚向心，寓意国歌凝聚亿万人民之心，发扬着无穷斗志与精神，"守护"建筑体量如双手环抱，象征为祖国人民筑成新的长城。"众心"建筑形体生长拥簇，体现中国人民万众一心，面对困难与艰辛，迎难而上的精神。

The National Anthem Museum is located in the south of the intersection of Dongba Line and Cultural Corridor, with a large difference in ground height and an irregular triangle. The total land area is 12,683. 22m², The longest distance from east to west is about 130m, and the longest distance from north to south is 160m. The design of the National Anthem Museum embodies the spiritual connotation of "The March of Volunteers": the "Condensation" of the architectural shape implies that the national anthem unites the hearts of hundreds of millions of people, carries forward the infinite fighting will and spirit, and "guards" of the building are like embracing arms. They symbolize the construction of a new Great Wall for the people of the motherland. "All Hearts" of the building shape grows and embraces together, reflecting the spirit of united mind and courage of Chinese people to face difficulties and hardship.

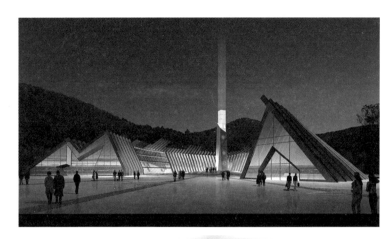

日月新天
New heaven

韶山创新成果专题展示馆
Shaoshan Innovation Achievements Special exhibition Hall

"千里来寻故地"，抚今追昔，豪情满怀，百年前毛主席奋笔写下《水调歌头·重上井冈山》；"可上九天揽月，可下五洋捉鳖，谈笑凯歌还。世上无难事，只要肯登攀。"是中华儿女的凌云壮志和对未来的无限神往。随着科技的发展，"可上九天揽月，可下五洋捉鳖"的探索宇宙航天梦和探测深海大洋梦，在今天的中国已全部变为现实。设计构思为"凝心聚力，三航筑梦"，航空、航天、航海三馆以"人"字形布局向中心齐聚，拱起顶部环形的创新成果展厅，寓意着拥抱星辰大海，凝聚中国力量，托起伟大梦想；大写的"人"字，体现了中国人民的伟大团结精神。充分利用场地高差，将建筑嵌入土地之中，使其底层通过阶梯式绿化广场和草坡隐藏起来，塑造破土而出、展翼腾飞的气势，体现了伟大梦想精神。韶山创新成果专题展示馆将开放、包容、共享的姿态赋予此地更加丰富的场地精神，续写"惟楚有材"的辉煌胜景。这里不仅是科技的历史汇聚点，也是不同人群和文化活动的空间汇聚点，更是韶山的过去、现在、未来的交汇之处。

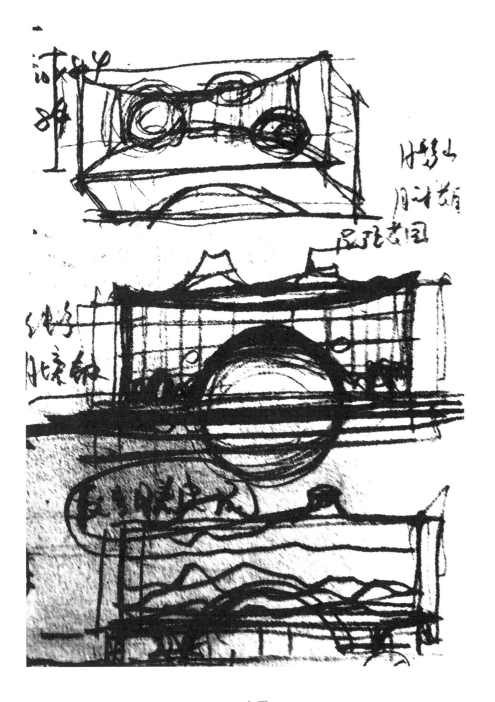

小暑

薰风愠解引新凉，小暑神清夏日长。
断续蝉声传远树，呢喃燕语倚雕梁。

——清·乔远炳《夏日》

韶山创新成果专题展示馆

1. 缘起

韶山，伟大领袖毛主席的故乡，坐拥年游客量2000万人次的红色旅游客群，旅游基数大，过夜率却极低，如何从传统的红色观光游过渡到新型体验式旅游，从"欢迎游客来"到"让游客留下来"是韶山旅游应该思考的问题。1965年，毛泽东同志在诗词中用"可上九天揽月，可下五洋捉鳖"描绘了探索宇宙的航天梦和探测深海的海洋梦。经过几代人的努力，嫦娥五号已携带月球样品回家。2020年12月，中国国家航天局正式宣布把湖南韶山作为月壤异地容灾备份存储基地，以告慰主席"可上九天揽月"的夙愿，面对月壤落户韶山这一重大机遇，韶山科技创新小镇应运而生。

2. 规划定位

韶山科技创新小镇背靠景区、面临城区，区位优势明显。总规划用地面积约1278666.7m²（1918亩），其中建设用地约400000m²（600亩），计划分三期实施。依托举世珍稀的"月壤"资源和主席故乡为两大中心点，聚焦红色传承、成果展示、科普教育、互动体验、文化旅游，项目定位为展示我国重大科技创新成果的重要窗口，开展爱国主义教育、研学实践和创新精神培训的重要阵地，促进红色文旅融合的重要载体。韶山创新成果专题展示馆为其一期工程，规划用地70266.7m²（105.4亩），将建设以"久有凌云志"为主题的航空馆，以"可上九天揽月"为主题的航天馆，以"可下五洋捉鳖"为主题的航海馆，以"世上无难事，只要肯登攀"为主题的"中国第一"创新成果馆和湖湘创新馆，打造集教育、展示、科普、旅游等多功能于一体的核心场馆。

3. 设计思路

设计围绕3个问题——如何表达韶山红色基底与科技创新的文化主题和精神内涵，凸显项目的独特性与地标性？如何协调周边环境及韶山城市整体功能的关系？如何打造场馆内部空间形态，使建筑与展陈有机融合，塑造具有创新性、体验性、互动性的展览空间？从内容出发，梳理展陈脉络，协调空间关系，通过对空间意向的把握，表现毛主席的气势和整体的意境，同时融入创新建筑技术，使展馆成为可持续发展典范。打造一个外部精神的丰碑，通过内部创新的路径，同时实现全域经济的提升和创新科技的发展。

4. 设计立意

设计构思为"凝心聚力，三航筑梦"，航空、航天、航海三馆以"人"字形布局向中心齐聚，拱起顶部环形的创新成果展厅，寓意着拥抱星辰大海，凝聚中国力量，托起伟大梦想，体现了中国人民的伟大团结精神。顶部极具科技感的环形展厅，营造出悬浮的未来感，象征着"巡天遥看一千河"的潇洒自在和"世上无难事、只要肯登攀"的伟大奋斗精神。将中国第一颗人造地球卫星"东方红一号"作为连接韶山红色文化与科技文化、连接梦想与未来的切入点，以"东方红一号"轨道倾角68.44°，将三个场馆沿中心线旋转布置，顶部创新成果馆内环近似"东方红一号"的运行轨迹，设计一个真实比例的"东方红一号"模型装置沿椭圆轨道运行；外环直径83m，象征毛主席83年的峥嵘岁月，既传承了红色基因，也展现出一代又一代中国科技工作者和中国人民的伟大创造精神。设计充分利用场地高差，将建筑嵌入土地之中，使其底层通过阶梯式绿化广场和草坡隐藏起来，塑造破土而出、展翼腾飞的气势，体现中国人民的伟大梦想精神。

5. 建筑设计

设计手法：设计将建筑分散布局，环形底部架空，透过建筑中间，可形成一条与山景融合的视线廊道。整体建筑布局形态与山体边缘相契合，底层覆土似山体的延伸，消减了建筑高度，使建筑与天鹅山周边环境和谐，藏而不露。

展陈思路：通过底层的公共空间连接三个分馆，形成一个总序厅"科技创新看中国"，同时将以航空探索、航空器物展示为核心的凌云馆，以向海图强、深海探测为主题的五洋馆，和以航天科普、月壤展示及互动体验为主题的九天馆，进行三馆串联，形成既分区明确又互相串联的流线组织。

功能布局：根据主题场馆特性，考虑不同场馆的展厅和公共空间，以集中连续的大型无柱展示空间和局部多层挑高的边庭，最大程度保证展陈空间的灵活性和利用率。在利用首层公共大厅组织整体游览动线的同时，将更多的服务性功能整合进来，通过特展互动、文创书吧、餐饮休闲、会议接待等多元化设计，将展览动线与体验休闲结合起来，为参观者带来更丰富互动体验，为后期运营提供保障。

可持续设计策略：建筑采用钢结构、装配式，利用地形现状将建筑下部覆土，3个场馆之间预留的架空通道与顶部圆环结合形成舒适的通风中庭，同时露天中庭设计下沉广场形成下凹式的雨水收集层，在屋顶和立面幕墙引入光伏发电材料，同时采用高效保温系统、智能空调系统等，实现绿色低碳建筑。同时通过大数据、云计算以及互联网等技术，场馆全业务实行信息化，打造创新的智慧展馆

Shaoshan Special Exhibition Hall of Innovation Achievements

1. Origin

Shaoshan, the hometown of the great leader Chairman Mao, has an annual number of 20 million red-tourism visitors. The tourist base is large, but the overnight visitor rate is extremely low. "Attracting visitors" and "letting them stay" is a question that Shaoshan tourism should think about. In 1965, Comrade Mao Zedong described the spaceflight dream of exploring the universe and the dream of exploring deep sea in his poems as "go up to the Heaven to fetch the moon, and go down to the ocean to catch turtles". After efforts of several generations, Chang'e 5 returned home with lunar samples. In December 2020, the China National Space Administration officially announced that Shaoshan, Hu'nan Province would be used as a remote disaster backup point for lunar soil sample to comfort the chairman's long-cherished wish of "fetching the moon". Upon the great opportunity of lunar soil sample being cherished in Shaoshan, Shaoshan Technology Innovation Town came into being.

2. Planning orientation

Shaoshan Science and Technology Innovation Town, backed by scenic spots, faces urban areas with obvious location advantages. The total planned land area is over 127 hectares, including about 40 hectares of land for construction, which is planned to be implemented in three phases. Relying on the world's rare "moon soil "resources and Chairman Mao's hometown as two central points, it focuses on revolutionary inheritance, achievement display, popular science education, interactive experience and cultural tourism. The project is positioned as an important window to display major scientific and technological innovation achievements of our country and develop patriotism. It is an important destination for education, research practice and innovative spirit training, and an important carrier to promote the integration of revolutionary culture and tourism. Shaoshan Innovation Achievement Special Exhibition Hall is the first phase of the project, with a planned land area of over 7 hectares . It will build an aviation hall with the theme of "Ling Yunzhi", an aerospace hall with the theme of "Go to the heaven to fetch the moon", and a navigation hall with the theme of "Catch Turtles in the Ocean", the "China's First" Innovation Achievement Hall and the Hu'nan Innovation Hall with the theme of "where there's a will there's a way", to create a multi-functional venue featuring the integration of education, exhibition, popular science, tourism and so on.

3. Design ideas

The design revolves around three questions, i.e. how to express the cultural theme and spiritual connotation of the revolutionary base and technological innovation in Shaoshan and highlight the uniqueness and landmark of the project? How to coordinate the relationship between the surrounding environment and the overall function of the city of Shaoshan ? How to create the interior spatial form of the venue, make the architecture and exhibition organically integrate, and create an innovative, experiential and interactive exhibition space? Starting from the content, we sort out the exhibition context, coordinate the spatial relationship, and bring out Chairman Mao's momentum and overall artistic conception through the grasp of the spatial intention, while incorporating innovative architectural technologies, make the pavilion a model of sustainable development. It creates a monument of external spirit, through the path of internal innovation, while realizing the improvement of the global economy and the development of innovative technology.

4. The design concept

The design concept is marked by "concentrate the strength, build the dream with three lines". Three halls of aviation, aerospace, and navigation gather in the center in the shape of "people". The Innovation Achievement Exhibition Hall with an arched top symbolizes the great heaven which summons together the Chinese strength, holds up the great dream, and embodies the great spirit of unity of the Chinese people. The ring-shaped exhibition hall with a strong sense of technology at the top creates a sense of the future suspended, symbolizing the unrestrained freedom of "scanning the sky and watching a thousand rivers" and the great fighting spirit of "where there's a will there's a way". Taking China's first artificial earth satellite "Dongfanghong-1" as the entry point to connect revolutionary culture and technological culture of Shaoshan, as well as connect dreams and the future. These three halls were arranged in a rotary pattern with an angle of 68.44 degrees similar to Dongfanghong-1 satellite, making the inter circle of the Innovation Achievement Exhibition Hall seemingly the orbit, and a real-scale model device of "Dongfanghong No. 1" is designed to run along an elliptical track. The diameter of the outer ring is 83 meters, symbolizing Chairman Mao's 83 years of efforts. The glorious years has not only inherited the revolutionary gene, but also demonstrated the great creative spirit of Chinese scientific and technological workers from generation to generation and the Chinese people. The design makes full use of the height difference of the site, embeds the building into the land, and hides the ground floor through the stepped green square and grassy slope, bringing about a momentum of breaking out of the ground and spreading its wings, reflecting the great dream spirit of the Chinese people.

5. Architectural design

Design methods: The design disperses the layout of the building, and the bottom of the ring is overhead. Through the middle of the building, a sightseeing pathway merged with the mountain can be formed. The overall architectural layout conforms to the edge of the mountain, and the ground floor is covered with soil like the extension of the mountain, which reduces the height of the building and makes the building harmonious with the surrounding environment of Swan Mountain, hidden but not exposed.

Combination of exhibition ideas: Connect these three branch halls through the public space on the ground floor, forming a general hall entitled "science and technology innovation of China". And, at the same time, the Lingyun Hall which is centered on aviation exploration and display of aviation equipment, Wuyang Pavilion with the theme of deep-sea exploration and Jiutian Pavilion with the theme of aerospace science popularization, lunar soil display and interactive experience are connected to form a streamline organization with clear divisions and connection.

Functional layout: According to the characteristics of the theme venues, we consider the exhibition halls and public spaces of different venues, and focus on the continuous large-scale column-free exhibition space and the partial multi-storey lifted side-courtyard to maximize the flexibility and utilization of the exhibition space. While using the public hall on the first floor to organize the overall tour routes, it integrates more service functions. Through diversified designs such as special exhibition interaction, z cultural and creative book-bar, catering and leisure and conference reception, the exhibition movement and experience are integrated. The combination of exhibition and leisure will bring a richer interactive experience for visitors and provide guarantee for the later operation.

Sustainable design strategy: The building adopts steel structure and prefabricated type. The existing terrain is used to cover the lower part of the building. The overhead passage reserved among these three venues is combined with the top ring to provide comfortable ventilation. At the same time, the open-air atrium is designed with a sunken square to collect rainwater. Photovoltaic power generation materials are applied to the roof and facade curtain wall, and a high-efficiency thermal insulation system and an intelligent air-conditioning system are installed to realize a green and low-carbon building. At the same time, through technologies such as big data, cloud computing and the Internet, the entire business of the venue is informatized to create an innovative smart pavilion eco-system.

韶山创新成果专题展示馆

Shaoshan Innovation Achievements Special exhibition Hall

用地面积：
70290.88m²
建筑面积：
38748.23m²
容积率：
0.45
设计时间：
2021-2022
建造时间：
方案

项目位于韶山市清溪镇天鹅山，湘宁线以西，韶山大道以南。地块总体为南高北低，现场地标高在85.28~100.90m 间，用地形状呈多块台地，高差较大，设计需对高差进行处理，尽量减少土方量，节约造价。根据地形及周边城市道路标高情况，场地内部标高基本上按照道路标高控制，以确保地面雨水能迅速排入城市道路排水系统中，且城市道路雨水不至于流入基地内。

将建设以"久有凌云志"为主题的航空馆，以"可上九天揽月"为主题的航天馆，以"可下五洋捉鳖"为主题的航海馆，以"世上无难事，只要肯登攀"为主题的"中国第一"创新成果馆和湖湘创新馆，打造集教育、展示、科普、旅游等多功能于一体的核心场馆。

（下图：韶山创新成果专题展示馆总图）

This project is located in Swan Mountain, Qingxi Town, the City of Shaoshan, west of Xiangning Line and south of Shaoshan Avenue. The plot is generally high in the south and low in the north. The site elevation is between 85.28-100.90m, and the land is of multiple platforms. There is a great height difference. The design needs to deal with the height difference to minimize the amount of earthwork and reduce the cost. According to the terrain and the elevation of surrounding urban roads, the internal elevation of the site is basically controlled according to the road elevation, to ensure that the ground rainwater can be quickly diverted into the urban drainage system so that the urban road rainwater will not flow into the site base. It will build an aviation hall with the theme of "Ling Yunzhi", an aerospace hall with the theme of "go to the heaven to fetch the moon", and an navigation hall with the theme of "Catch turtles in the ocean", the "China's First" Innovation Achievement Hall and Hu'nan Innovation Hall with the theme of "Where there's a will, there's a way", to create a multi-functional venue featuring the integration of education, exhibition, popular science, tourism and so on.

（Below: General plan of Shaoshan Special Exhibition Hall of Innovation Achievements）

Site Area：
70,290.88m²
CFA：
38,748.23m²
FAR：
0.45
Design Time：
2021-2022
Construction Time：
Program

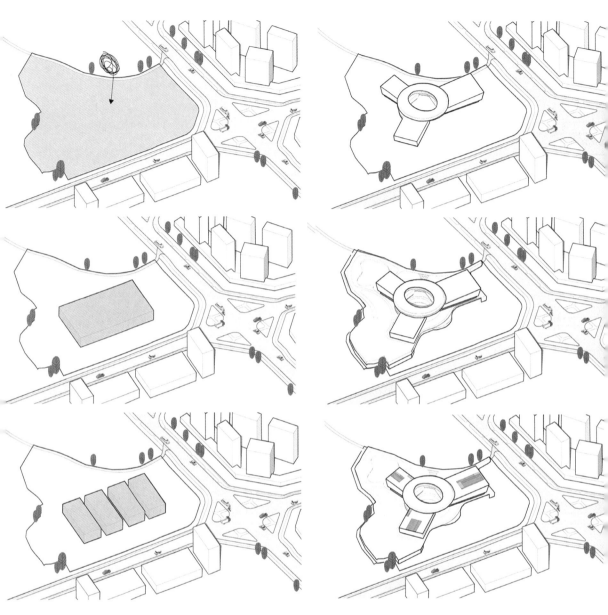

Block Generation

1. The project faces the city, favored with obvious regional advantages.

2. Analyze the original volume of the building.

3. Break down the volume, and determine the area of each exhibition hall.

4. Concentrate the strength, build the dream with three lines.

5. Take advantage of the existing height difference, set up a semi-basement, broaden the street-side square, and increase the living room area for the city.

6. Increase the natural lighting of the building and install photovoltaic power generation panels on the roof.

体块生成

1. 项目面临城区，区位优势明显。

2. 分析建筑原始体量。

3. 分散体量，确定各个展览馆面积。

4. 凝心聚力 三航筑梦。

5. 利用现有高差，设置一个半地下室，并且打开临街广场面，增加城市客厅。

6. 增加建筑自然采光面，屋顶铺设光伏发电板。

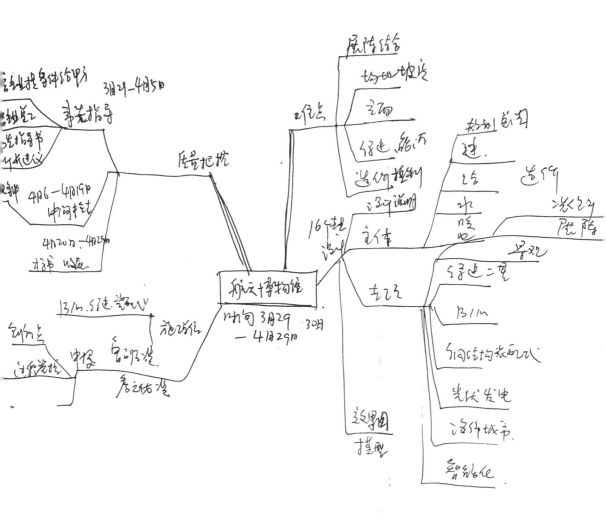

The ring-shaped exhibition hall with a strong sense of technology at the top creates a sense of the future suspended, symbolizing the unrestrained freedom of "scanning the sky and watching a thousand rivers" and the great fighting spirit of "where there's a will, there's a way".

顶部极具科技感的环形展厅，营造出悬浮的未来感，象征着"巡天遥看一千河"的潇洒自在，"欲与天公试比高"的豪迈自信和"世上无难事、只要肯登攀"的伟大奋斗精神。

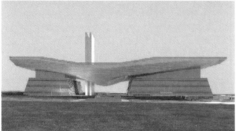

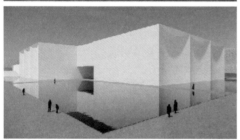

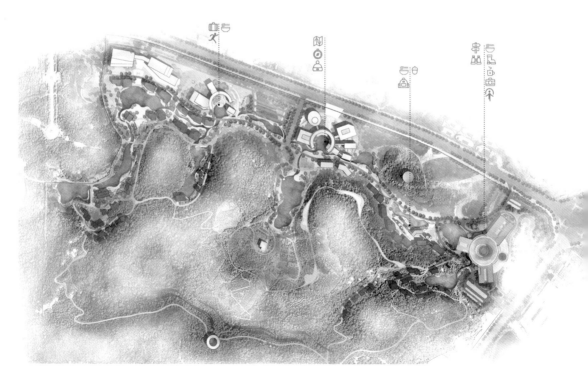

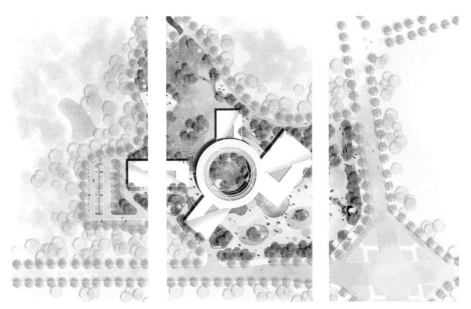

建筑采用分散布局方式和尺度控制，与周边天鹅山自然环境有机融合，布局形态与山体边缘相契合。环形底部架空，透过建筑中间，可形成从狮子山、韶山大道与如意路交叉口、建筑中庭、至天鹅山山顶的一条视线廊道，将建筑群与山体环境共融。

The building adopts a decentralized layout and scale control, which organically integrate themselves with the surrounding natural environment of Swan Mountain, and the layout form fits with the mountain edge. Through the middle of the building, a visual corridor can be formed from Lion Mountain, the intersection of Shaoshan Avenue and Ruyi Road, the atrium of the building, to the top of Swan Mountain, integrating the architectural complex with the mountain environment.

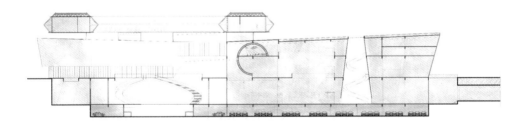

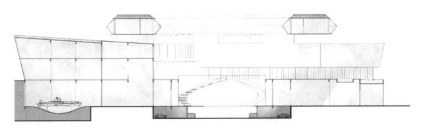

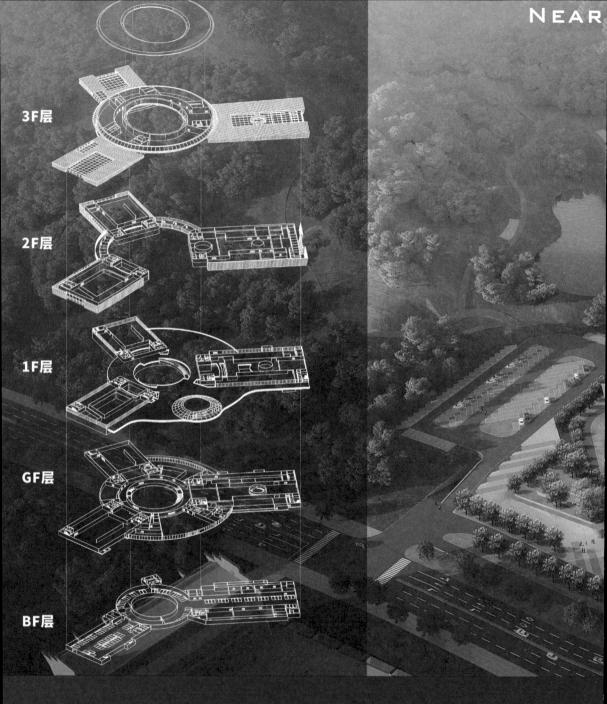

3F层

2F层

1F层

GF层

BF层

将中国第一颗人造地球卫星"东方红一号"作为连接韶山红色文化与科技文化、连接梦想与未来的切入点。以"东方红一号"轨道倾角
68.44°,将三个场馆沿中心线旋转布置,顶部创新成果馆内环近似"东方红一号"的运行轨迹,将设计一个"东方红一号"真实比例、直径
1m 的模型装置沿椭圆轨道运行;外环直径 83m,象征毛主席 83 年的峥嵘岁月,既传承了红色基因,也展现了一代又一代中国科技工作者
和中国人民的伟大创造精神。

Taking China's first artificial earth satellite "Dongfanghong-1" as the entry point to connect revolutionary culture and technological culture in Shaoshan, as well as connecting dreams and the future, these three halls were arranged in a rotary pattern with an angle of 68.44° similar to Dongfanghong — 1 satellite, making the inter circle of the Innovation Achievement Exhibition Hall seemingly the orbit. A real-scale one-meter-diameter model device of "Dongfanghong No. 1" is designed to run along an elliptical track; the diameter of the outer ring is 83 meters, symbolizing Chairman Mao's 83 years of efforts. The glorious years has not only inherited revolutionary gene, but also demonstrated the great creative spirit of Chinese scientific and technological workers from generation to generation and the Chinese people.

Optimize the facades. The architectural style, featuring the theme of Shaoshan culture and technology, is more agreeable with the environment. The design of the first floor is relatively transparent, the entrance hall goes harmoniously with the circular design of the top, which is both simple and reeks of a sense of science and technology. Meanwhile, the round dome to the public lobby is used to collect light. These three venues are designed with light gray stone facades, which provide a overall low-profile and elegant style. The facade of the top UFO-like exhibition hall elevation is simplified, which is composed of three concentric circular decorative components. Its internal design is single-floor exhibition hall which is relatively simple for construction and also full of sense of technology and practicality. The ground floor of the building is combined with the landscape, telling the story of Shaoshan through detailed landscape design and reflecting the revolutionary base of Shaoshan with local colors.

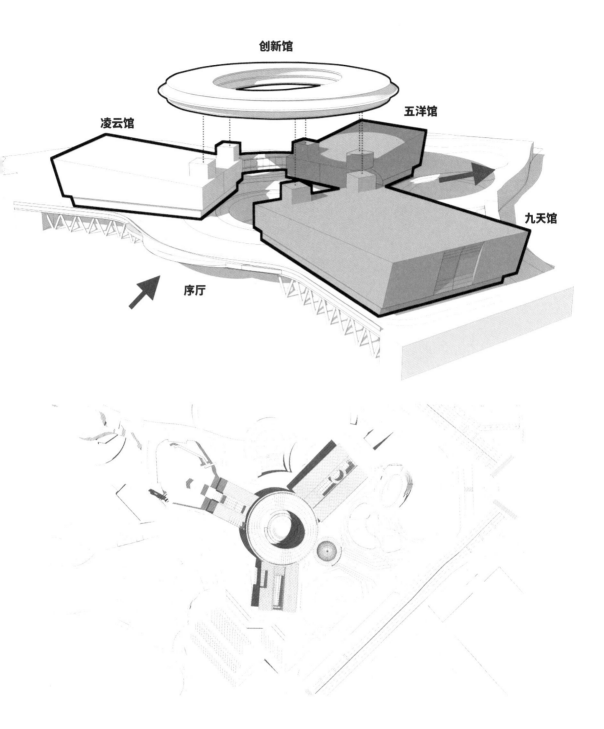

创新馆

凌云馆

五洋馆

九天馆

序厅

优化建筑立面，建筑风格凸显韶山文化和科技主题，与环境更加协调。首层设计相对通透，入口门厅设计呼应顶部圆环，具有科技感又相对简洁，同时在公共大厅顶部形成圆形采光顶，三馆采用浅灰色石材立面，整体低调素雅，顶部圆环展厅立面简化，形似飞碟，由 3 个同心圆装饰构件叠加而成，内部为单层展厅施工较简便，兼顾科技感和实用性。建筑底层与景观结合，通过景观的细部设计讲出韶山故事，用局部色彩体现韶山红色基底。

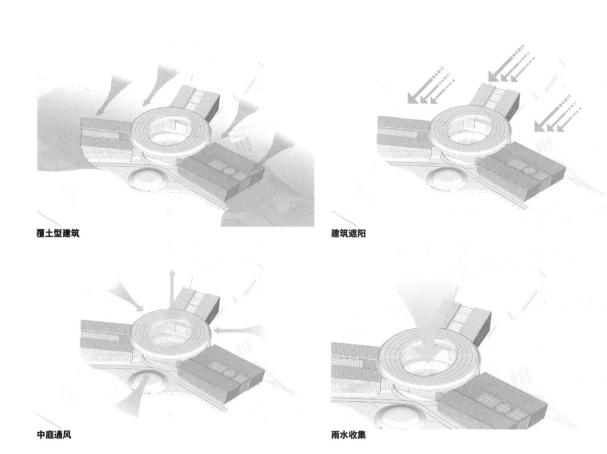

覆土型建筑

建筑遮阳

中庭通风

雨水收集

光伏发电

中庭

透水

雨水收集

景观集水池

回收利用　中水池　压力水池　储水池　初期滞留

污水管道　溢流管道

绿建措施

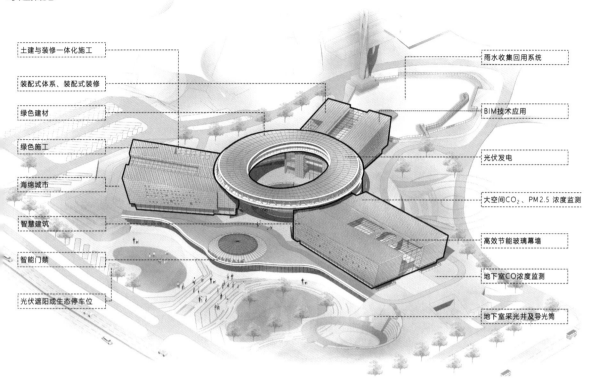

土建与装修一体化施工

装配式体系、装配式装修

绿色建材

绿色施工

海绵城市

智慧建筑

智能门禁

光伏遮阳或生态停车位

雨水收集回用系统

BIM技术应用

光伏发电

大空间CO_2、PM 2.5 浓度监测

高效节能玻璃幕墙

地下室CO浓度监测

地下室采光井及导光筒

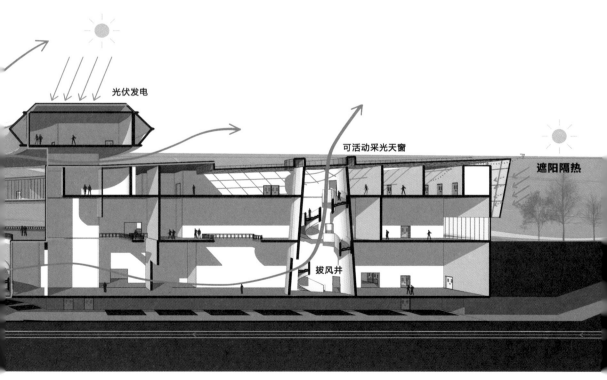

光伏发电

可活动采光天窗

遮阳隔热

拔风井

韶山创新成果专题展示馆
Shaoshan Innovatial exhibition Hall

"Travelling thousands of miles to find the old place", looking back on the past, and was full of pride, Chairman Mao assiduously wrote "Shuidiaogetou-Going up to Jinggang Mountain Again" , which goes:"Go to the heaven and embrace the moon, dive into the ocean and catch turtles. Laugh and sing out aloud. Impossibility means nothing to a strong will." This song has been inspiring Chinese descendants for generations to cherish great ambitions for the future.

With the development of science and technology, "Embracing the moon and catching turtles" has all become a reality for modern China. With an open, inclusive, and sharing attitude, Shaoshan Aerospace Science and Technology Town will endue the place with more abundant site spirit and continue the glorious scenery of "only chu has talent" It is not only the historical gathering place of science and technology, but also the spatial gathering place of different people and cultural activities, as well as the intersection of Shaoshan's past, present and future.

"千里来寻故地"，抚今追昔，豪情满怀，毛主席奋笔写下《水调歌头·重上井冈山》。"可上九天揽月，可下五洋捉鳖，谈笑凯歌还。世上无难事，只要肯登攀。"是百年前中华儿女的凌云壮志和对未来的无限神往。

随着科技的发展，"可上九天揽月，可下五洋捉鳖"，在今天的中国已全部变为现实。韶山航天科技小镇将以开放、包容、共享的姿态赋予此地更加丰富的场地精神，续写"惟楚有材"的辉煌胜景。这里不仅是科技的历史汇聚点，也是不同人群和文化活动的空间汇聚点，更是韶山的过去、现在、未来的交汇之处。

观物取象
Observing things like

农村住宅与民宿设计
Design of rural houses and homestays

设计是一场对话，在静默中观望着过去与未来攀谈，与时间窃窃私语。消融了实有的空间隔阂，在建筑的有无之间，回归原初的自然，才有了最本真的自我，收放自如，肆意洒脱。曲折迂回的行进动线，可任意游走、观赏，拐角处形成挡墙、山石，还有零零星星洒落到墙上斑驳的树影，更有了些天地间极净的自然意味。在平和中俯身自观，自观空间的虚实，亦观内心的缩影，唤醒观照内心的力量，相由心生，亦随心灭。每一个触点，每一个接口，每一条接缝的存在都是为了强调构想中的作品的静谧气质。每一个完成的、完善的作品都有一种魅力，就仿佛我们沉浸于这高度完善的建筑实体的魅力一样。我们的注意力或许是被某个细节俘虏，例如地板及墙壁上摇曳的树影，它们逆着阳光随意摆弄姿态，任由时光流逝而描绘出一幕幕独一无二的黑白灰画面。这些东西令我们感动，它们甚至超越了标志和符号，开放而空寂。

大暑

大暑三秋近，林钟九夏移。
桂轮开子夜，萤火照空时。

——唐·元稹《咏廿四节气诗·大暑六月中》

隐宿

建筑的更新与介入带来了新旧建筑的思考与重塑

Hidden Residence

The renewal and intervention of the building have brought about the thinking and remodeling of the old and the new buildings

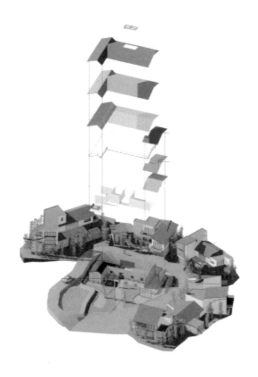

建筑布局看似偶然得来，实则是慎思之后的结果。在天然环境之中事实上有无数隐藏的线，岸线、溪线、脊线、坡线、坳线……形成了一张无形的网络，而建筑与场地设计要想自然地融入这个网络，其设计中隐含的轴线关系必须要和这张自然的网络相契合，看似随意的摆布实则具有非常精巧的对位关系，无形的线之间产生了无形的张力，并进一步产生了无形的场，场中气韵流动，亦聚亦散，人与环境自然地合而为一，一脉共生。

The architectural layout seem random, but it is actually the result of careful thinking. In fact there are countless hidden lines in the natural environment. The coastline, the stream line, the ridge line, the slope line, and the terrain line form an invisible network. If architecture an site design want to be naturally integrated into this network the implicit axis relationship in the design must be consisten with this natural network. The seemingly random arrangemen actually has a very exquisite counterpoint relationship, an invisible tension is generate among invisible lines, furthe producing an invisible field, i which the charm flows, gather and disperses, and people and th environment naturally becom one and coexist in one vein.

210

Remote Mountains in Tengchong

The renewal and intervention of the building have brought about the thinking and remodeling of the old and the new buildings

腾冲远山

腾冲远山

建筑的更新与介入带来了新旧建筑的思考与重塑

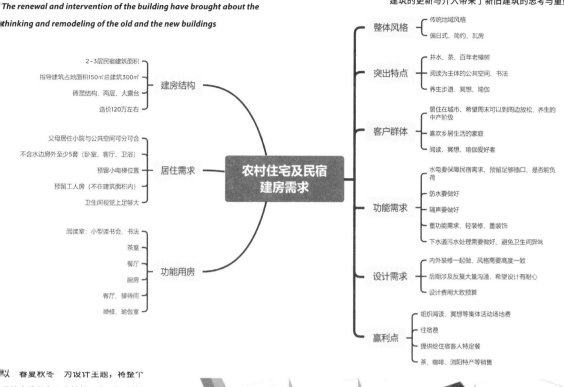

整体风格
- 传统地域风格
- 偏日式、简约、瓦房

突出特点
- 井水、茶、百年老樟树
- 阅读为主体的公共空间、书法
- 养生步道、冥想、瑜伽

客户群体
- 居住在城市、希望周末可以到周边放松、养生的中产阶级
- 喜欢乡居生活的家庭
- 阅读、冥想、瑜伽爱好者

功能需求
- 水电要保障民宿需求、预留足够插口、是否能负荷
- 防水要做好
- 隔声要做好
- 重功能需求、轻装修、重装饰
- 下水道污水处理需要做好、避免卫生间异味

设计需求
- 内外装修一起做、风格需要高度一致
- 后期涉及反复大量沟通、希望设计有耐心
- 设计费用大致预算

赢利点
- 组织阅读、冥想等集体活动场地费
- 住宿费
- 提供给住宿客人特定餐
- 茶、咖啡、浏阳特产等销售

建房结构
- 2~3层民宿建筑面积
- 指导建筑占地面积150㎡总建筑300㎡
- 砖混结构、两层、大露台
- 造价120万左右

居住需求
- 父母居住小院与公共空间可分可合
- 不含水边房外至少5套（卧室、客厅、卫浴）
- 预留小电梯位置
- 预留工人房（不在建筑面积内）
- 卫生间视觉上足够大

功能用房
- 阅读室：小型读书会、书法
- 茶室
- 餐厅
- 厨房
- 客厅、接待间
- 禅修、瑜伽室

农村住宅及民宿建房需求

以"春夏秋冬"为设计主题，将整个场地大致分为四个地块，春、夏、秋、冬依次引入四个地块中。最前端的区域为餐厅、大堂等公共活动区域，逐步过渡到客房、禅室等最安静的区域。而春夏秋冬四季则由繁密茂盛过渡到寂静萧瑟，相互对应，并通过材料、植物的不同来进一步表达主题。

Taking "Spring, Summer, Autumn, and Winter" as the design theme, the whole site is roughly divided into four plots, and the Spring, Summer, Autumn, and Winter are introduced into the four plots in turn. The most front-end areas are public activity areas such as restaurants and lobbies, gradually shifting to the quietest areas such as guest rooms and meditation rooms. Four seasons of Spring, Summer, Autumn and Winter transit from denseness and lushness to quietness and bleakness, which correspond to one another, further expressing the theme through different materials and plants.

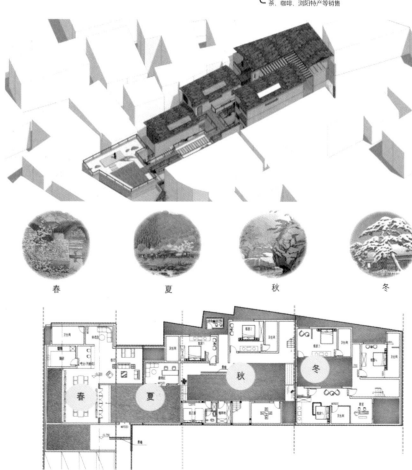

春　　　夏　　　秋　　　冬

筑·巢
聚落再生

Build a Nest
Community Regeneration

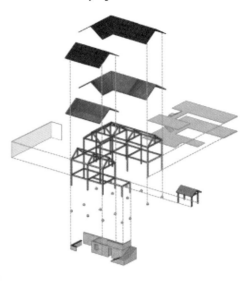

设计分析了湘西山区现有的场地环境，建筑材料结构体系，空间形态，建筑设备系统和设计建造智慧等建筑要素，以更新迭代和可适性的角度为出发点，探索既保留传统元素又具备现代的适用性，既环保节能又绿色低碳性，既经济又有收入的开放性。

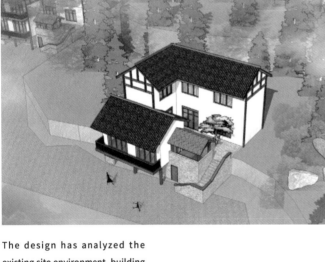

The design has analyzed the existing site environment, building materials, the structural system, the spatial form, the building equipment system, design and construction ideas, and other building elements in the mountainous area of west Hu'nan Province. From the perspective of renewal iteration and adaptability, it explores the green and low-carbon nature of preserving tradition and modern applicability, environmental protection and energy conservation, and the openness of economy and income.

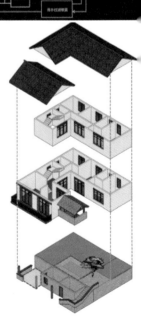

山云吞吐翠微中，淡绿深青一万重，
比景只应天上有，岂知身在妙高峰。

The mountains nestle among the clouds, whilst decorated with layers of greenery in varied shades. The scene ought to be from the Heaven, yet who's to know it is right by us.

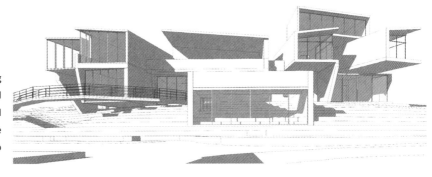

Change

《尔雅》：妃、合、会，对也

《说文解字》：运也。从车专声。知恋切

风带走了

焦虑的过程

世界依然自我运转

结局有些许遗憾

春天会重启下一个梦想与欢喜

幸运不过是寂静的前夜

我们依旧热情的拥抱，放下、反思

清零、自在

过程就是答案

屡败屡战

让眼中的星星

互相点亮

汇聚永恒的光华

交融与流动

Blend and flow

郴州市公共中心配套区概念性规划设计
Conceptual Planning and Design of the Supporting Area of the Public Center of Chenzhou

静观空间里面的建筑和外面的景观，把自然的思考融入每一栋建筑中，不破坏原有的景观加以创新，我们选择摒弃建筑的重量感和永恒性，而追求像水一样随时代变化的流动性。正如伊东丰雄所说："建筑是鲜活的，如同流水一般，建筑和我们的人生都在不断地发展变化"。"流动"的想法来源于中国古代习俗"曲水流觞"，古代文人们通过曲水流觞饮酒赋诗，是一种具有文化性的游戏，正如创作的过程就像不断地从自然之河中汲取灵感之水，富于流动性的建筑单体各具独立空间，又交融于空间的波纹中。由于物体、时空的介入，建筑便被赋予了生机和使命，建筑本身虽然是静止的，却并不妨碍人们用它来感觉流动。尊重城市的整体环境，从整体格局中提炼空间要素，确定城市的控制原则，确立区域的总体视觉架构，将城市空间的形象要素在建筑、景观层面上加以整合。通过对城市空间节点，景观核心，中央景观轴，标志性建筑单体的营造，与周边景观相结合，创造鲜明的城东新区商业中心形象。我们将进行城市风貌、空间形态、建筑风貌等要素的控制和详细的空间节点设计，最终构筑融入自然的公共中心配套区。

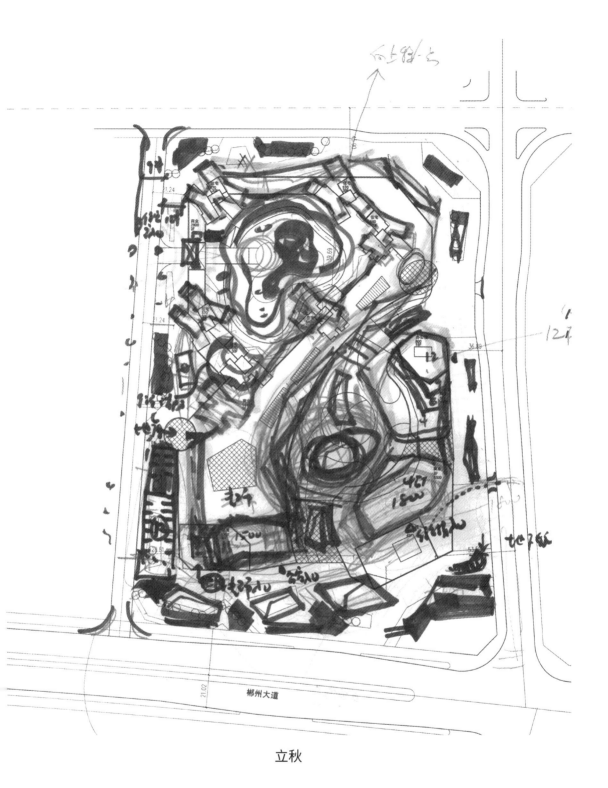

立秋

乳鸦啼散玉屏空，一枕新凉一扇风。
睡起秋色无觅处，满阶梧桐月明中。

——宋·刘翰《立秋》

郴州市公共中心配套区概念性规划设计

Conceptual Planning and Design of the Supporting Area of the Public Center of Chenzhou

用地面积：
75075.13m²
建筑面积：
423500m²
容积率：
4.5
设计时间：
2013-2014

方案以矿晶体为设计构思，把形象元素融入设计之中，立面造型简洁大气，高雅华贵。建筑单体如一根根纯净的水晶矗立在城东新区之中，建筑单体通过中心旋转裙楼相连，仿佛矿晶体般，拥簇融合、和谐共生。

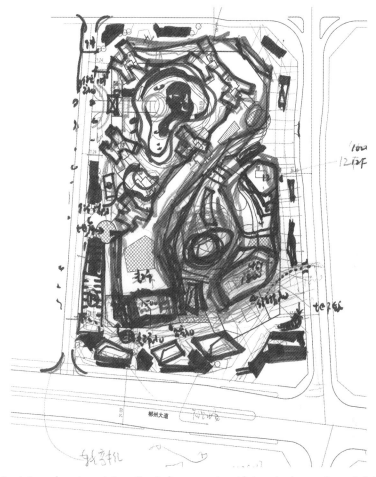

Site Area：
75,075.13m²
GFA：
423,500m²
FAR：
4.5
Design Time：
2013-2014

The plan takes mineral crystals as the design concept and integrates image elements into the design. The facade is simple and elegant, noble and luxurious. The building monomers stand like pure crystals in the East New District. They are connected by the central rotating podium, like mineral crystals, clustering, merging and coexisting in harmony.

We also incorporate the element of water and design the podium into the shape of flowing water with continuous twists and turns, flexible and soft. The design combines the landscape to connect the residential and commercial waterscapes so that the water would flow around in the building, adding a sense of fun and color to the whole center. Meanwhile, the smooth commercial image also echoes with the exhibition center in the west.

我们也融入水的意向，将商业裙房设计成流水的造型，连绵曲折，灵活柔美，设计结合景观，把住宅和商业的水景联系在一起，让涓涓细流在建筑中汇聚，为整个中心增添一份趣味和色彩。同时，流畅的商业形象也与西侧会展中心遥相呼应。

节约与友好
Economical and friendly

长沙市两型社会示范项目
Changsha Two Types of Social Demonstration Project

2007 年 12 月，一颗绿色种子播撒进湖南这方内陆高地改革沃土，长株潭城市群被国务院批准为全国资源节约型、环境友好型社会建设综合配套改革试验区，长沙作为省会城市担负起为国家探索两型路径的重大责任，奋力追逐光荣与梦想。春华秋实，岁物丰成。十三年来，长沙从"两型"荣光走来，向着幸福的深处走去，两型社会建设已从盆景变风景、化苗圃为森林，成为全市干部群众内化于心、外化于行的自觉行动，呈现出神形兼备、丰盈充实的全域化格局，为加快建设富饶美丽幸福新湖南，奋力谱写新时代中国特色社会主义湖南新篇章提供"绿色动能"。三湘小区是长沙市旧城改造的示范项目，附近是中南林业科技大学和民政学院，通过前期策划与详细分析，将城市康养与创新孵化相结合，实证了城市更新与生态转型的设计结合。

处暑

天地乾坤始渐肃，鹰隼捕鸟稷乃登。
冷热交换试拳脚，一场秋雨一场寒。

——当代·贵谷子《二十四节气之处暑》

227

长沙市两型社会示范项目

Changsha Two Types of Social Demonstration Project

用地面积：
82769.09m²
建筑面积：
458425m²
容积率：
4.57
设计时间：
2017-2018

规划设计以三湘客车制造厂、三湘小区的历史记忆为载体。以第四代住宅为亮点，打造商业、居住、历史记忆展示及多种商业配套服务等功能于一体的综合性两型示范商业街区。

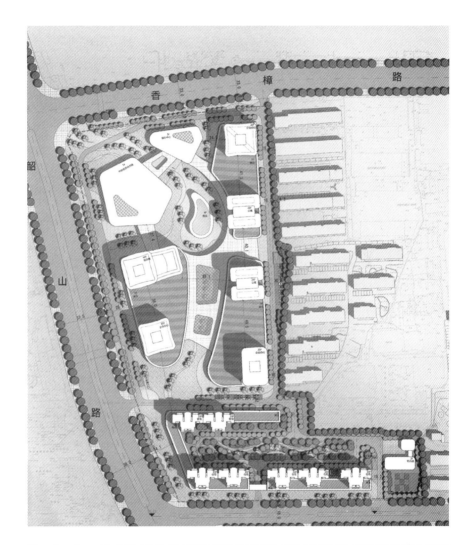

Site Area：
82,769.09m²
GFA：
45,8425m²
FAR：
4.57
Design Time：
2017- 2018

The planning and design take the historical memory of Sanxiang Bus Manufacturing Plant and Sanxiang Community as the carrier. With the fourth-generation residence highlighted, it will create a comprehensive two-type demonstration commercial street with such functions as commercial, residential, historical memory display, and various commercial supporting services.

Planning and design goals:

Create a spatial form with a pleasant scale and a regional symbol.

Diverse industrial functions are mixed to enhance the vitality of the region.

Highlight regional characteristics and lead the cultural highland.

Characteristic urban ecological environment, highlighting two types of demonstrations.

Build scenario space and increase regional viscosity.

A diverse and convenient transportation system.

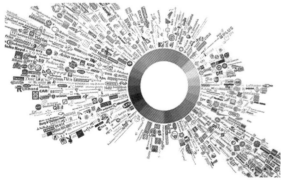

规划设计目标：

塑造尺度宜人、有地域标志的空间形态。

多样产业功能混合，提高地区活力。

突出地区特色，引领文化高地。

城市特色生态环境，凸显两型示范。

建设情景空间，提高地区黏度。

多样的便捷交通体系。

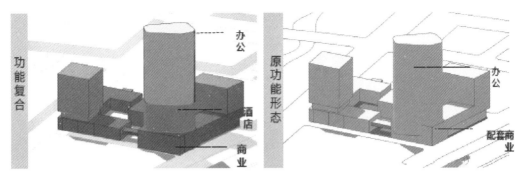

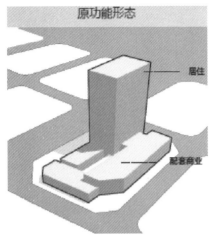

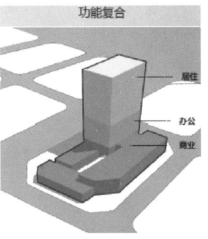

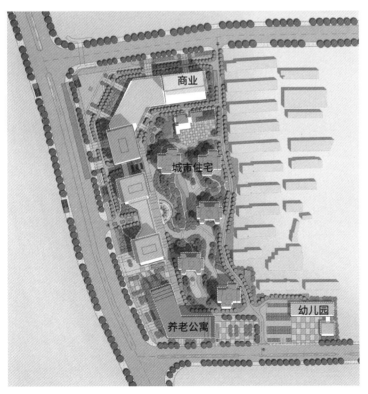

传承与转译
Inheritance and translation

张家界武陵源民族文化广场
Wulingyuan Mingzu Cultural Square

建筑是城市文化主要的，也可以说是最重要的载体。场地位于湖南省张家界武陵源区，是大型旅游文化演出实体魅力湘西剧场的扩建工程。因此，对城市传统文化的体现，是设计的一个重要诉求。我们力求超越在视觉和符号层面对历史的浅薄再现，把传统的建筑形式提炼、抽象，再以现代的建筑语言、材料来表现人们记忆中的对湘西传统建筑屋顶这一形式的映像，在精神层面继承历史文脉。通过对湘西传统建筑形式的表达诉求，按照合乎现代理性的建筑生成原则，与建筑的功能需求，建造逻辑，结构、设备技术的要求等，紧密结合，以现代建筑的设计方法、语境再现传统的精神和神韵。

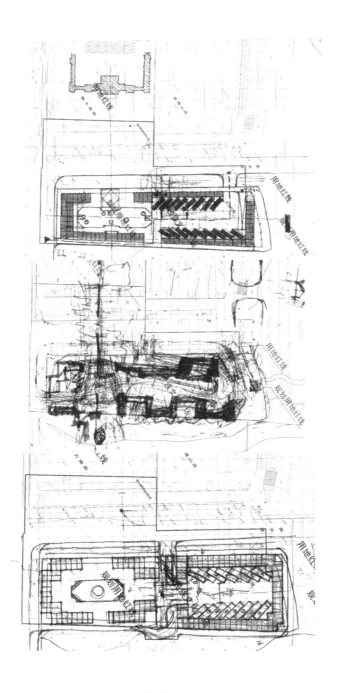

白露

白露觅觅秋分起，又见丰收稻晚米。

——清·《二十四节气歌》

张家界武陵源民族文化广场

Wulingyuan Mingzu Cultural Square

用地面积：
15946.5m²
建筑面积：
21131.14m²
容 积 率：
0.39
设计时间：
2014-2015
建造时间：
2015-2017

世界自然遗产武陵源民族文化广场位于湖南省张家界武陵源景区核心地段高云路，与蜚声海内外的大型旅游文化演出实体魅力湘西剧场隔路相互辉映，项目两面临水，项目现状用地地势平坦，东侧临河 200m，北侧临河 71m，临高云路绿化带 187m，项目整体地块形状为临高云路长条形；两侧临河、临桥；桥面高出地块约 1.5m；项目土地 11180m²（16.77 亩）与周边防洪堤面积 4766.7m²（7.15 亩），总面积含河滩共 19866m²（29.8 亩）进行一体规划。

（下图：武陵源游客服务中心）

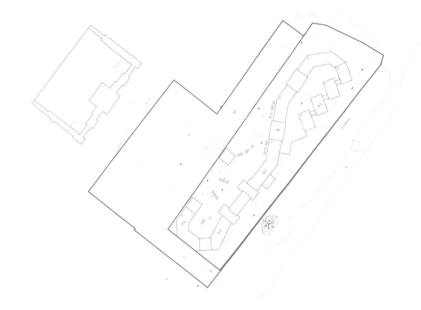

Site Area：
15,946.5m²
CFA：
21,131.14m²
FAR：
0.39
Design Time：
2014-2015
Construction Time：
2015-2017

Wulingyuan National Cultural Square as the World Natural Heritage is located on Gaoyun Road, the core area of Wulingyuan District, Zhangjiajie, Hu'nan Province. It is only one street away from the large-scale world-renowned tourism and cultural performance entity Theatre of Xiangxi Charm. The project is adjacent to the river on two sides. The terrain of the project is flat, 200m from the river in the east, 71m from the river in the north, and 187m from the green belt of Gaoyun Road. The overall shape of the project is a long strip along Gaoyun Road; the river and bridge are on both sides. It is about 1.5m higher than the plot. With 16.77 acres of land and over 0.5 hectare of flood dam, a total of 23.92 acres (total area of 29.8 acres including the river beach) is planned for the project.

（Below: Wulingyuan Tourist Service Center）

设计从多角度全方位探讨方案的各种可能性，力求平衡各方面的需求，设计出最优化方案。

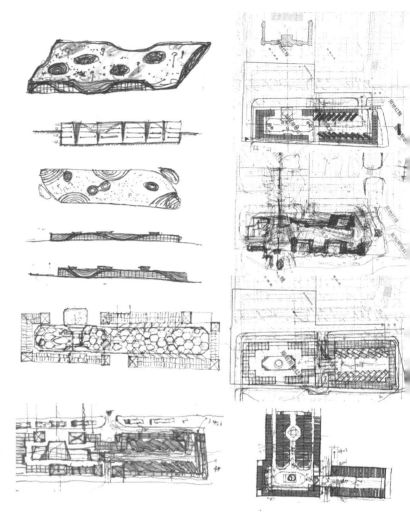

We discuss the feasibility of the design plan from multiple perspectives, strive to balance the needs of all aspects, and design a optimal plan.

我们的灵感来自于湘西传统建筑文化，对各种建筑元素特点进行分析，然后用当代的建筑语言和方式重新塑造我们的项目。

首先我们被当地的传统形式的屋顶所启发，连绵不断的屋顶形成一条起伏运动的天际线，然后我们根据抽象的概念设计出了一个复杂的新物体。

We are inspired by the traditional architectural culture of western Hu'nan Province, analyze the characteristics of various architectur
elements, and then reshape our project with contemporary architectural language and methods.

First, we were inspired by the local traditional form of the roof. The continuous roofs formed a wavy skyline, and then we designec
complex new object based on the abstract concept.

承载与使命
Load bearing and Mission

汨罗人民医院
Miluo People's Hospital

"建筑物之创造，为社会解决衣食住行中住的问题"，可以说建筑物是文化的纪录者，是历史之反照镜。在我们看来，建筑可阅读，设计始终有温度。在汨罗人民医院设计之初，设计的核心思想是从一个物质的建筑回归人本，希望通过设计创作回馈他人、回馈社会，当为了人民生活更美好来进行创造时，这件事就变得有意义了。社会责任感，源于对国家、对民族的认同感，更源于一种使命。医院的使命是救死扶伤，设计师的使命是将社会需求与社会结合，使这种承载生死的"最复杂建筑"拥有一些记忆的温度和厚度。

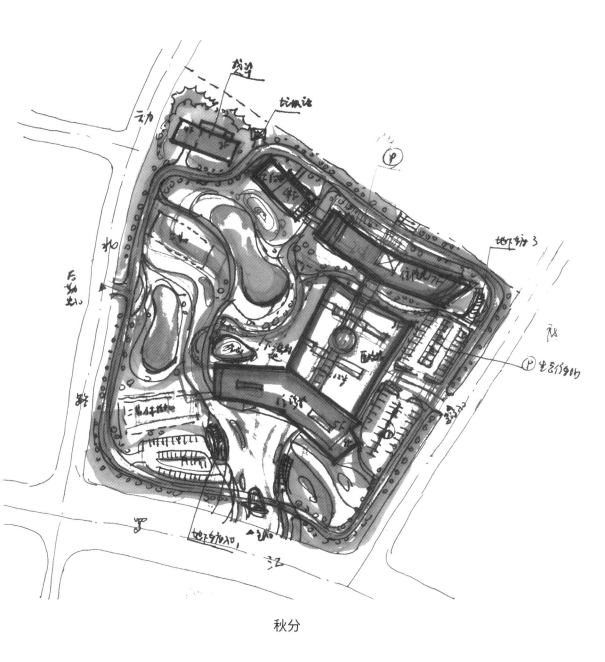

秋分

一分秋意一分凉，野外繁露披衣裳。

八九菊黄蟹儿肥，风和气爽丹桂香。

——当代·贵谷子《二十四节气之秋分》

汨罗人民医院

Miluo People's Hospital

用地面积：
100560m²
建筑面积：
119681.07m²
容 积 率：
1.2
设计时间：
2018

项目基地东临规划路，南看汨罗江大道，西依劳动北路，北靠玉池路。场地呈倒梯形，内部现状地形较为平坦，南北长约 330m，东西宽 280~360m。基地北侧为两块预留发展用地，临近汨罗江，景观资源优越。按使用方的意愿拟设计成集医疗、教学、科研为一体的现代化三甲综合性非盈利医院。设计在满足管理需求下也应尽可能保证学科和共享中心的连接高效性。

（下图：汨罗人民医院总图）

Site Area：
100,560m²
GFA：
119,681.07m²
FAR：
1.2
Design Time：
2018

The project is bordered by Planning Road in the east, Miluojiang Avenue in the south, Laodong North Road in the west, and Yuchi Road in the north. The site is in the shape of an inverted trapezoid, and the current internal terrain is relatively flat. It is about 330m long from north to south and 280 to 360m wide from east to west. On the north side of the base are two pieces of reserved development land, close to the Miluo River, with superior landscape resources. According to the wishes of the user, it is planned to be designed as a modern Third-Grade First-Class comprehensive non-profit hospital integrating medical treatment, teaching and scientific research. The design should also ensure the efficiency of the connection between the discipline and the sharing center as much as possible while meeting the management needs.

(Below: A general map of Miluo People's Hospital)

The design concept:

The word"mi" means the sun by the water, and it implies the place where life and hope rise. The graphic design of the hospital coincides with the image of"mi". Based on the understanding of the traditional dragon-boat culture of"The Dragon Boat Festival by the Miluo River", we use a soft and elegant arc to connect the buildings together. The overall architectural layout presents a dynamic posture from east to west, which contains a profound meaning of "The Life Boat, Healing Green Valley".

"Healing Green Valley".The green valley presents two images: the vigorous green and the curvature of the mountain shape give it a gentle side. At the same time, the valley embraced by the mountains reminds people of the protection of life. These two images are metaphors for the characteristics of the hospital, i.e. people-oriented, care for life, and the valley of healing.

设计理念：

"汨"字是水边的太阳，也寓意着生命和希望升起的地方。医院的平面设计暗合"汨"字字形象。基于对传统"汨罗江畔端午习俗"龙舟文化的理解，我们采用柔畅飘逸的弧线，一气呵成地将各个建筑串联在一起，总体建筑布局自东向西呈现出动感的姿态，蕴含着"生命之舟，治愈绿谷"的深刻含义。

"治愈绿谷"，绿色的山谷呈现出两种形象：生机盎然的绿色以及山形的弧度赋予其柔美的一面。同时，群山对谷地的环抱又使人联想到对生命的保护。这两种形象是对医院特性的隐喻：以人为本，呵护生命，治愈之谷。

The layout guides the flow of people and forms an efficient transportation system through the horizontal medical street and streets vertical to the medical street which has better guidance and efficiency. The design is flexible, and the functions of each module can be replaced with each other to meet various needs of the hospital's future development. The modular layout facilitates the uniformity of various specifications of the entire hospital, greatly reducing the difficulty of construction.

The hospital encompasses all kinds of people and operates throughout the day. It will have an impact on the space and transportation of the city. We will make positive impacts on urban connections, resource application, and urban appearance through design to promote urban development.

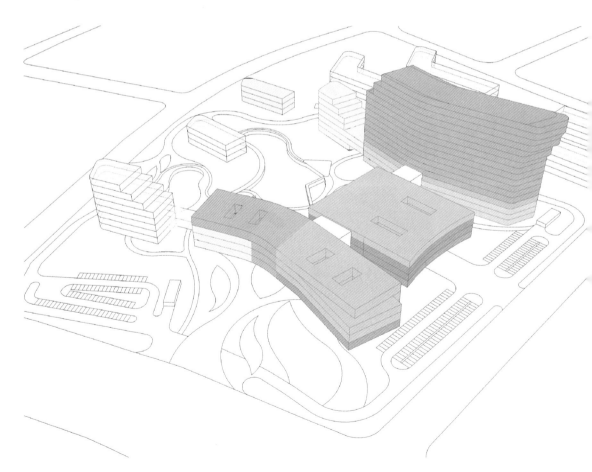

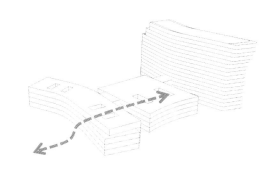

导向性骨架引导人流，通过水平向医疗街及医疗街上的垂直交通，形成高效的骨架交通体系，具有更好的导向性和效率。
设计具有灵活性，各模块的功能可以相互置换，以应对医院未来发展的各种需求。模块化的布局方便整个院区建筑的各项规格统一，大大降低了施工难度。

医院包含了各类人群，并24小时运作。会对城市的空间、交通产生影响，我们通过设计让城市连接、资源互用，从城市界面上做出积极影响，促进城市生长。

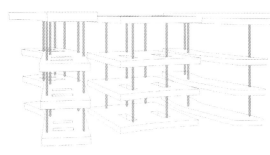

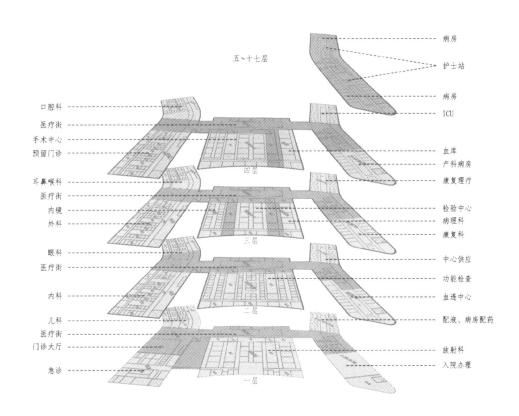

聚合与散落
Aggregation and scattering

洋湖湿地公园商业综合体
A Commercial Complex of Yanghu Wetland Park

生活与自然的关系，就像青山与绿水，明月与松风，让人平和、畅神。悉心留住自然的温度，生活自然会真心相待。心有山水相依，生活重回自然。为让洋湖湿地公园成为一件和自然相融相生的美学作品，我们与大自然合作，最大化激发出自然的潜力。利用周边的景观与形式简单的建筑精妙地组合在一起，建筑风格与区域环境协调、融合。

寒露

红叶深秋凝景象，蝉噤荷残偶见霜。

晨早洎塘腾雾气，袅袅轻盈舞逸上。

——当代·贵谷子《二十四节气之寒露》

洋湖湿地公园一期 D 区

Phase I Area D of Yanghu Wetland Park

用地面积：
113600m²
建筑面积：
50414m²
容 积 率：
0.27
设计时间：
2012

规划布局最大限度利用湘江、湿地公园景观，突出"街道""院落"主题，把形式简单的建筑精妙地组合在一起，形成街区、庭院。使置身其间的人享受游玩湿地公园后愉悦心情的延伸。建筑风格注重与区域环境湖湘文化的协调、融合。景观设计以湿地公园已建园林景观为蓝本，即是湿地公园景观的一部分，又是湿地公园至湘江的过渡段。

（下图：洋湖湿地公园一期 D 区总图）

Site Area：
113,600m²
GFA：
50,414m²
FAR：
0.27
Design Time：
2012

The plan layout takes full advantage of the scenery of Xiangjiang River and the wetland park. It brings out the theme of "road" and "yard", and exquisitely integrates simple buildings together to form street blocks and yards. One would be entitled to more fun after visiting the wetland park. The architectural style features the unity with the regional environment with Huxiang culture. The landscape design is based on the existing garden landscape of the wetland park, which is a part of the wetland park landscape and the transition section from the wetland park to Xiangjiang River

(Below: A master plan of Area D of Phase 1 of Huyang Wetland Park)

自然的秩序
Order of nature

北山天乙书院
Beishan Tianyi College

天，中华文化信仰体系的一个核心，狭义仅指与地相对的天；广泛意义上的天，即道、大自然、宇宙。乙是天干的次位。其在方向上指东方，五行属木，为阴。天亦可表示植物的生长周期，乙指幼苗或出芽。天乙教育寓意以自然之道培育人才幼苗。书院建筑提倡"礼乐相成"的文化意境。 一是体现出了"礼"的秩序规范，建筑布局中轴对称来体现空间的层次感和秩序感，区分建筑场所的主次与尊卑。二是喻意着"乐"的和谐统一，建筑群的庭院和天井空间在空间维度和时间维度上丰富了建筑群的层次，使轴线建筑体的意义更加饱满，创造亲近自然、朴素典雅的人文气息场所和严谨的建筑布局轴线相互渗透，与场地周边的自然山水交相呼应，与书院形成对立统一 的和谐整体。

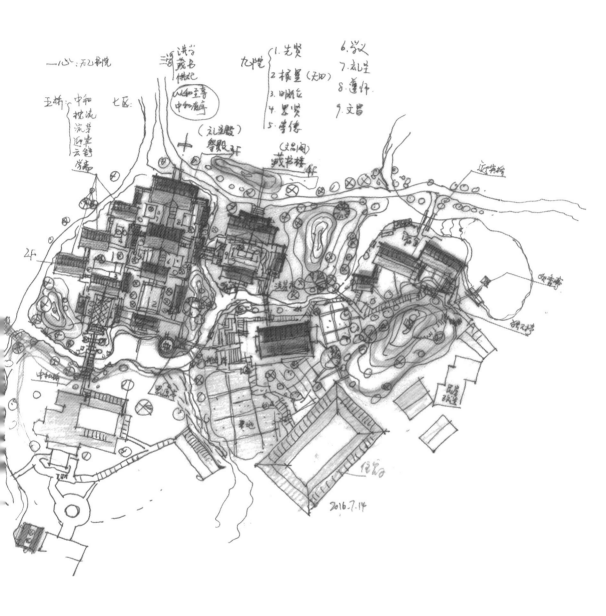

霜降

气肃霜降渐冷凉，草木枯萎凋零黄。
月夜秋云没落水，总有青绿独自芳。

——当代·贵谷子《二十四节气之霜降》

北山天乙书院

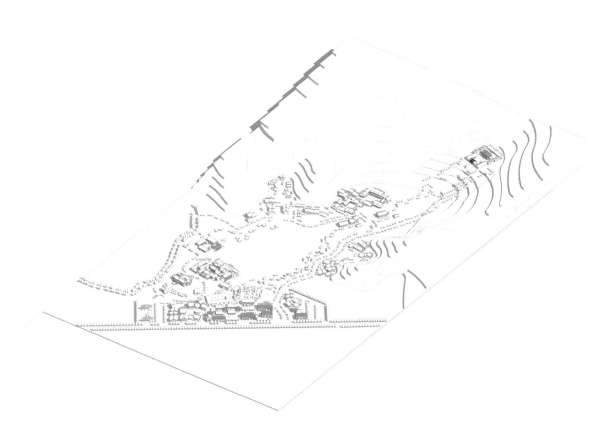

Beishan Tianyi College

用地面积：
153300m²
建筑面积：
20119.52m²
容 积 率：
0.11
设计时间：
2017-2018

项目总用地约为 153300m²（230 亩），主要为山地及农田，地形起伏较大，其中可建设用地为 40000m²（60 亩），除拟规划的书院区建设区域外，其余建设区域均为山地，地形起伏较大，如何在最大限度尊重自然的前提下结合地形地貌，充分发掘场地潜力，延续地形地势特色实现"传统民居村落型学校"的设计理念。

（下图：北山天乙书院总图）

Site Area：
153,300m²
GFA：
20,119.52m²
FAR：
0.11
Design Time：
2017-2018

The total land area of the project is over 153300m² , most of which is mountainous land and farmland. The terrain is undulating, of which 40000m² can be used for construction. Except for the planned construction area of the college area, the rest of the construction area is mountainous, and the terrain is undulating. How to bring the potential of the land to the fullest under the premise of maximum respect for nature and to fulfill the design concept of "a traditional residential village school" in line with the topographical features become a major topic.

(Below: General plan of Beishan Tianyi College)

它与城市若即若离，跨出门即万丈红尘，收回脚又可清风朗月。隐于北山，是厌弃了城市人的喧嚣浮华，却不舍人间浓浓的眷恋温情，即如陶渊明的桃花源，结庐人境，草屋鸡鸣，有良田美池桑竹之属。虽可得和穆清逸，却决不会像鲁迅先生所讥讽过的所谓隐士，泰山崩，黄河溢而目无见，耳无闻，依然心中有热血，有大爱。太平世界时可以采菊悠然，寄怀琴书。一旦时变，亦可拔剑而起，望义而归，担当天下。此北山之儒也。故北山有李默庵，有戴凤祥，有陈琼英，此皆隐而儒而侠之人也。

　　　　　　　　　　——王跃文

If it is far away from the city, it will be red dust when stepping out of the door, and it will be clear to the moon when you retract your feet. Hiding in Beishan is a renunciation of the hustle and bustle of urban people, but does not give up the strong nostalgia and warmth of the world, i.e. Tao Yuanming's peach blossom source, the land of humanity, thatched cottages, and the genus of Liangtian Meichi mulberry. Although available and Mu Qingyi, it will never be like the so-called hermit that Mr. Lu Xun once ridiculed, the Taishan landslide, the Yellow River overflows into the sight, the ears are unknown. There is still blood, and great love in the heart. In the peaceful world, you can pick chrysanthemums and place the feelings on instruments and books. Once the time changes, you can also pull up your sword, look back on the righteousness, and bear the day. This is also Beishan Confucianism. Therefore, there are Li Moan, Dai Fengxiang, and Chen Yingying in Beishan.

　　　　　　　　　　——Wang Yuewen

礼制下的建筑序列
通过中轴对称来体现空间层次感和秩序感，在同条
轴线中又层层递进、逐渐深入，区分建筑场所的主
次与尊卑，从而达到人精神上秩序的共鸣，予人心
理上关于礼仪和秩序的暗示。

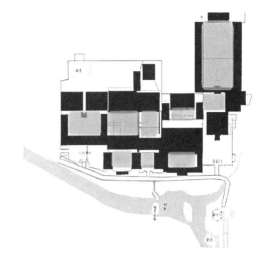

白鹿洞书院院落布局

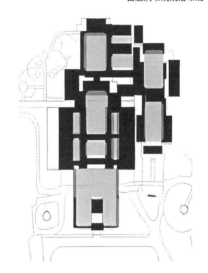

岳麓书院院落布局

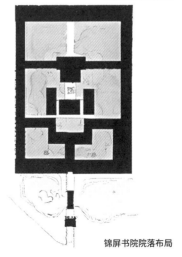

Building Sequence under Etiquettes
The central axis symmetry is used to reflect
the sense of layering and order of space. In
the same axis, it is progressively and gradually
deepened. It distinguishes the class and
integrity of the building place so as to achieve
the resonance of the spiritual order and to
imply a sense of etiquettes and order.

锦屏书院院落布局

乐制下的群体布局

书院建筑提倡"礼乐相成"的文化意境，除体现出
"礼"的秩序规范之外，还喻义出"乐"的和谐统一。
"乐"的精神主要体现在建筑群的庭院和天井空间上，
在空间维度和时间维度上丰富了建筑群体的层次，
使轴线建筑体的意义更加饱满，意在创造亲切自然、
朴素典雅的人文气息场所，和严谨的轴线建筑空间
互相交融渗透，与墙外的自然山林水系相呼应使
书院形成对立统一的和谐整体，在空间维度和时间
维度上丰富了建筑群体的层次，使轴线建筑体的意
义更加饱满，意在创造亲切自然、朴素典雅的人文
气息场所，和严谨的轴线建筑空间互相交融渗透，
与墙外的自然山林水系相得益彰使书院与自然环境
形成对立统一的和谐整体。

A group layout under the music system

The academy-building advocates the cultural
conception of"combination of courtesy and
music". In addition to embodying the order
and norms of courtesy, it also symbolizes the
harmonious unity of "music". The spirit of
"music" is mainly reflected in the courtyard
and patio space of the building group, which
enriches the level of the building group in
space and time dimensions, bringing more
meanings to the axis building, and it is intended
to create a friendly, natural, simple, and elegant
atmosphere that would ease the rigidness of
the axis building. The building space is blended
with each other and interacts with the natural
mountain forest system outside the wall to form
a unified and harmonious entity.

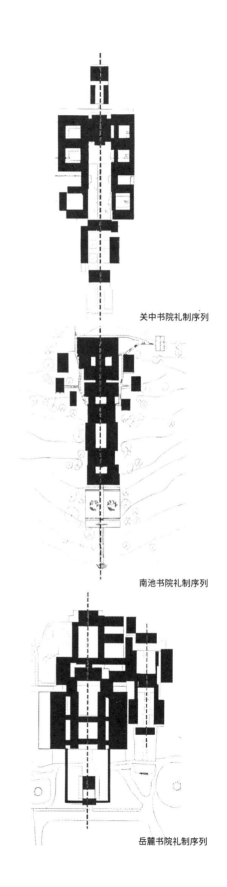

关中书院礼制序列

南池书院礼制序列

岳麓书院礼制序列

规划布局为："一心两带五区"即以北山书院为核心，以两条溪流水系为两带，围绕形成"书院区、运动区、生活区、耕读区、学院区"。

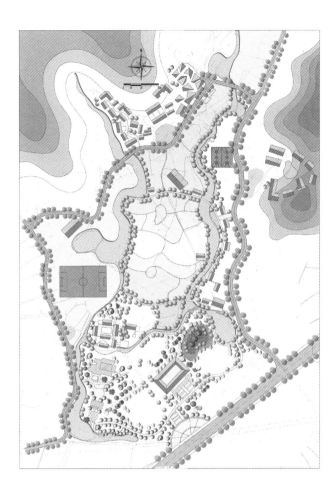

The planned layout is: "One Heart, Two Belts and Five Districts", i.e. with Beishan Academy as the core and two rivers as the two belts, forming " A School Area, a Sports Area, a Living Area , a Farming Area, and a College Area".

溪
流

溪
流

书院集培训，藏书等多功能为一体。采用院落围合式布局，
院落参考传统书院形式布置。

The academy integrates training, book-collection, and
other functions. The courtyard adopts an enclosed
layout which is arranged with reference to the
traditional academy form.

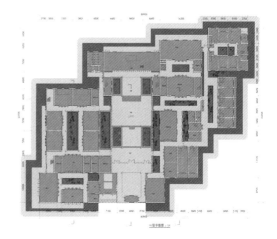

一层平面图

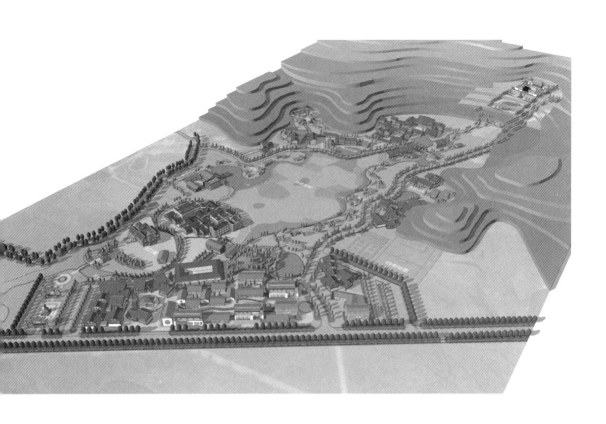

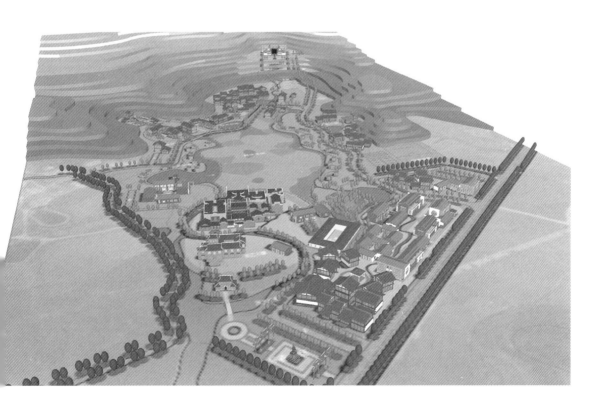

本源与和谐
Origin and harmony

天岳幕阜山国际度假旅游区
Tianyuemu Fushan International Resort

道家思想认为"道"是世界万物的本源，提倡道法自然，无为而治，与自然和谐相处。天岳幕阜山是道教圣地，中国道教传说中的仙居秘境洞天福地第二十五洞天，道教文化底蕴浓厚。素以山雄、崖险、谷幽、林秀、石奇、水美而著称，自然生态环境优美。我们通过挖掘天岳幕阜山的自然景观资源和文化特色，从适合现代旅游总体发展趋势出发，打造以山岳休闲、道教文化体验、农业、养老、休闲三大功能为核心的旅游休闲度假区。

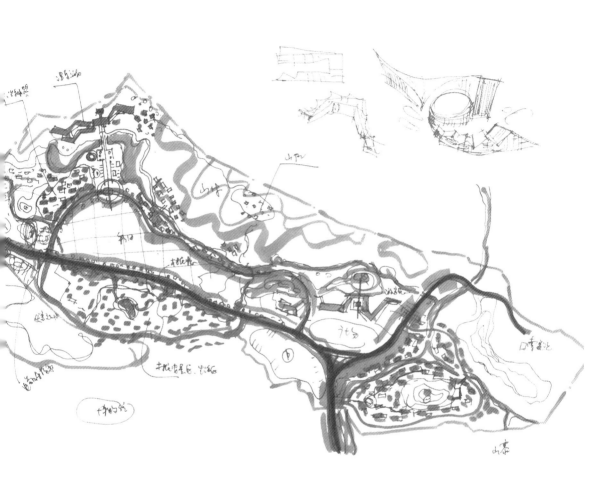

立冬

一年四季随岁走，今到立冬储寒衣。

农耕乃重盛德水，善修渠道莫闲田。

——当代·贵谷子《二十四节气之立冬》

天岳幕阜山国际度假旅游区修建性详细规划

Detailed Planning for Construction of Tianyue Mufushan International Resort Tourist Area

用地面积：
1700000m²
建筑面积：
13943.52m²
容积率：
0.008
设计时间：
2016

"天岳出、大道汇"
幕阜山为道教三十六洞天福地之第二十五洞天福地。设计围绕天极、天谷、天岳小镇等共计 25 洞天，打造群星环绕、璀璨生辉的旅游体系。
（下图：岳阳市君山区城市设计总图）

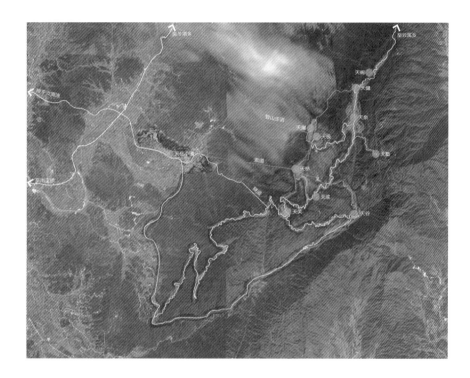

Site Area :
1,700,000m²
GFA :
13,943.52m²
FAR :
0.008
Design Time :
2016

"Tianyue, Avenue".
Mufu Mountain is the twenty-fifth one of the 36 Taoist resorts. A total of 25 resorts are designed around Tianji, Tiangu, and Tianyue towns, creating a brilliant tourist system.
(Below: A general plan of Junshan district, the city of Yueyang)

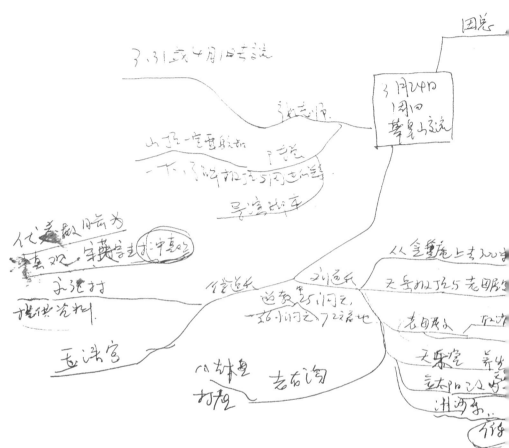

Based on the overall planning strategy of Pingjiang County, Mufu mountain tourist development vigorously develops the characteristic resources of Pingjiang County, integrates agriculture into tourist development, promotes the economic development of Pingjiang county with tourism, and realizes industry-based wealth for the people and tourism-based prosperity for the county. Under this strategic positioning, the planning area is divided into three major divisions:

Yongqiang village - Tianyue Tangquan health hot-spring vacation plate

Yifengjian sightseeing core area

Laolonggou outdoor sports valley

用地

道路

水系

基地

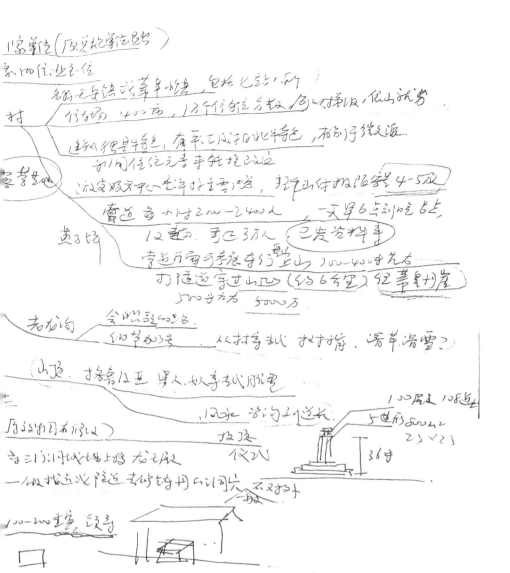

幕阜山旅游开发在平江县总体规划战略的基础上，大力开发平江特色资源，将农业纳入到旅游开发中来，以旅游业带动平江县经济发展，实现产业富民，旅游活县。并在此战略定位下将规划区域分为三个大板块：

永强村——天岳·汤泉健康温泉度假板块

一峰尖观光核心区

老龙沟户外运动谷

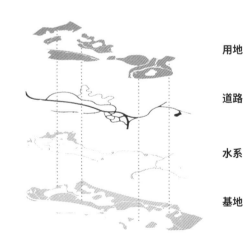

用地

道路

水系

基地

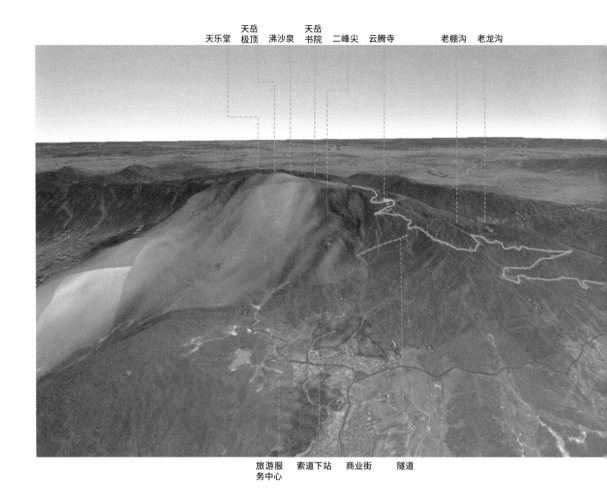

天乐堂　天岳极顶　沸沙泉　天岳书院　二峰尖　云腾寺　老棚沟　老龙沟

旅游服务中心　索道下站　商业街　隧道

Within the scope of this plan, land of agriculture and forestry is mainly used. There are some tourist reception facilities on the mountain, and there is a small amount of village residential land in Yongqiang Village. According to the characteristics of the base, it can be roughly divided into three independent blocks. The first block is the mountain-peak area, with an altitude of about 1,400-1,600m. The main attractions are Taiyuan Tiangong, Mufu Danya and Bosha Spring. The climate here is pleasant. The second block is the Laolonggou block, with an altitude of about 600-1,200m, featuring canyons, waterfalls, and forests, with a beautiful ecological environment. The third block is the Yongqiang village block, with an altitude of 220-300 meters. There are several scattered natural villages around, and most of the farmhouse buildings are modern and simple. There are a few characteristic buildings of northern Hu'nan residential houses. Most of the flat land at the mountain-foot is farmland, orchards and vegetable fields, and there are 2 water reservoirs.

规划范围内主要以农林用地为主，山上有部分旅游接待设施，永强村区块有少量村居住宅用地。依据基地特征大致可分为三个独立区块，第一区块为山顶区域，海拔1400~1600m，主要分布有太元天宫、幕阜丹崖、沸沙泉等主要景点，山顶空间开阔，气候宜人；第二块为老龙沟区块，海拔600~1200m，以峡谷、溪瀑、森林为景观特色，生态环境优美；第三块为永强村区块，海拔220~300m，有几个散布的自然村落，农居建筑大部分现代简约风格，有少量湘北民居特色建筑，山下平地大部分为农田、果园、菜地，并有2处水库。

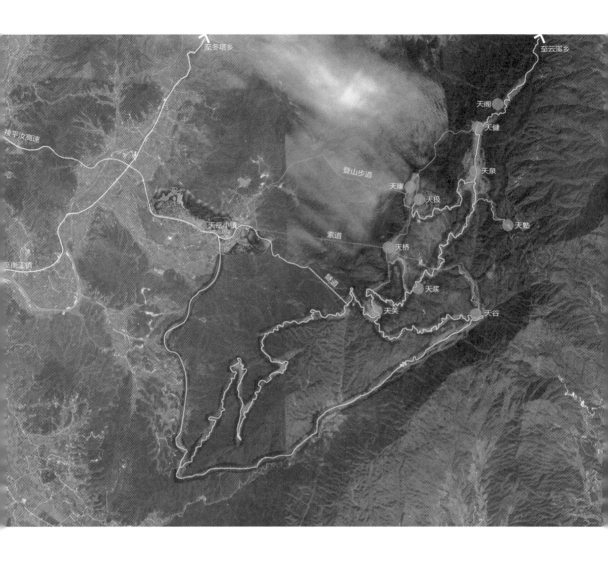

寻天问道，独与天地精神往来，打造国内云顶观光奇迹，道教文化示范体验基地，国家级森林生态景区，5A 级旅游景区，使一峰尖景群成为观光新"洞天"，在此寻天问道，体会自然和精神的统一。

Seeking the heaven and asking for the truth, interacting with the spirit of the heaven and earth alone, creating a domestic Yunding sightseeing miracle, a base of Taoist cultural demonstration experience, a national forest ecological scenic spot, and a 5A-level tourist attraction, making Yifeng Peak Scenic Group a new "sky-cave" for sightseeing. When visiting this place, you can experience the unity of the nature with spirit.

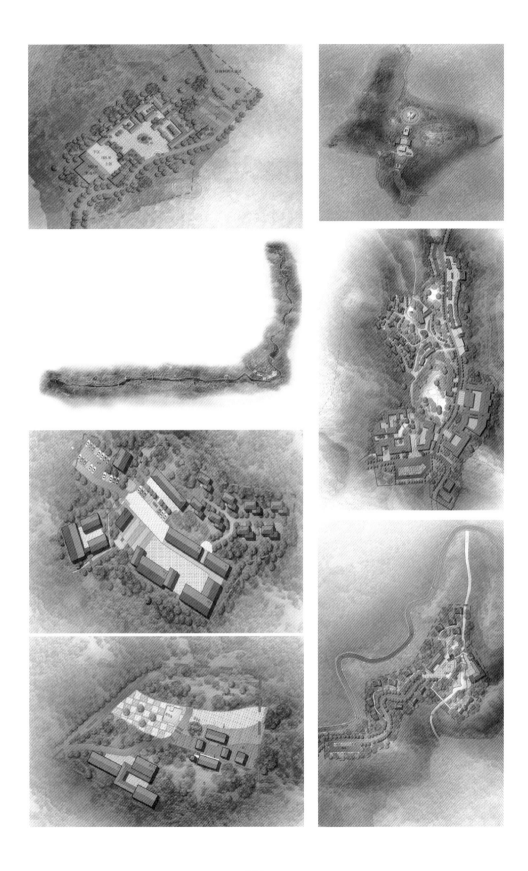

求是与创新
Seeking truth and innovation

四川广安小平干部学院
Sichuan Guang'an Xiaoping Cadre College

传承经典、突破创新

四川广安小平干部学校是一所以传承邓小平精神为核心和特色的全国一流党性教育基地和改革开放干部学院，设计者将邓小平"三落三起"的传奇经历融入到场地设计中，试图通过剖析邓小平光辉伟大的一生，向参与学习者传递邓小平在漫长的政治生涯中始终坚持思想解放、实事求是，敢说敢做、无私无畏的政治勇气，感受邓小平屡陷逆境、信念坚定的精神力量。学院整体采用"春天的故事，改革开放"为设计思路，秉承"传承经典，突破创新"的表现手法，以"方圆交融，刚柔并济"的表达方式呈现出来。把历史、文化与现代、自然串在一起，通过对传统的转译，基于现代建筑材料和建构手段，表达对地域文化的尊重；通过空间的围合与连接，与老校区在空间上形成联系，追求新旧建筑和谐共生的本质。

围合凝聚、庄重严谨

汲取川东传统院落"合而不闭"的营建智慧，追求和谐的审美理想。设计所追寻的凝聚感、向心力等意识形态，试图呼应邓小平团结一致的精神力量。通过中央的团结广场，形成富有凝聚力的校园中心和标志性交往空间，强化校园特色认知；通过建筑之间的组合连接，形成多层级交互式公共空间。建筑屋顶造型吸取传统川东地区民居的坡屋顶、跑马廊与现代建筑特色结合起来。

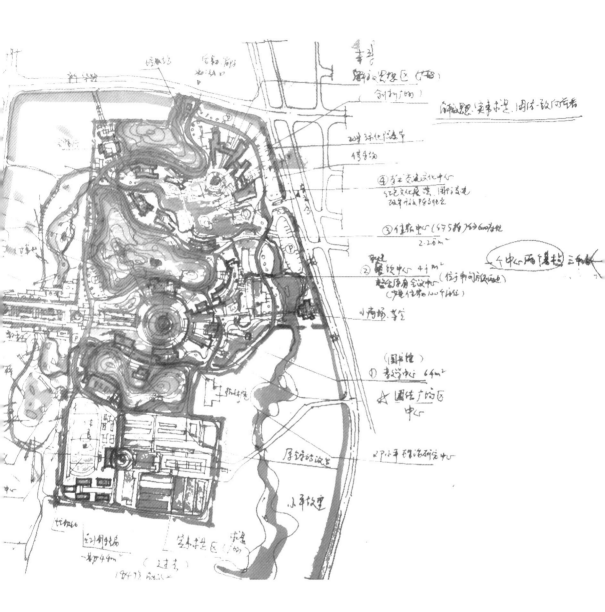

小雪

倒尽床头酒半罂，笑呼筇杖共闲行。

梅花照水为谁瘦，雪片倚风如许轻。

孟德遇冬思射猎，广文垂老谢才名。

归来跨火西窗下，独数城楼长短更。

——宋·陆游《闲步至鞠场值小雪》

四川广安小平干部学院

Sichuan Guang'an Xiaoping Cadre College

用地面积：
451882.26m²
建筑面积：
117552.40m²
容 积 率：
0.28
设计时间：
2020

四川广安小平干部学院位于广安市区北部，属于川渝合作区广安协兴生态文化旅游园区，基地东南侧紧邻邓小平故里景区，风景优美。西侧为协兴湿地公园，区域环境优美；东邻广花路，北抵新华路，西接内环路，交通条件良好。

在城市资源和优势整合下，未来本案定位为：以传承小平精神为核心和特色的全国一流干部党性教育基地和改革开放干部学院。本案功能定位为：以小平干部教育培训为主，兼具会议会展，国际交流，文化旅游为一体的开放城市新地标。

（下图：四川广安小平干部学院草图）

Site Area：
451,882.26m²
CFA：
117,552.40m²
FAR：
0.28
Design Time：
2020

Guang'an Xiaoping Cadre College in Sichuan Province is located in the northern part of the City of Guang'an, belonging to the Guang'an Xiexing Ecological and Cultural Tourism Park in the Sichuan- Chongqing Cooperation Zone. The southeast side of the base is adjacent to Deng Xiaoping's hometown scenic spot with beautiful sceneries. At the western side is Xiexing Wetland Park with a beautiful regional environment; adjacent to Guanghua Road in the east, Xinhua Road in the north, and Inner Ring Road in the west with good traffic conditions.

Under the integration of urban resources and advantages, the future of this project will be positioned as a national first-class cadre Party-education base and a reform and opening up cadre college with the core and characteristics of inheriting Xiaoping's spirit. The function of this project is positioned as a new landmark of an open city that focuses on the education and training of Xiaoping-type cadres, combining conferences and exhibitions, international exchanges and cultural tourism.

(Below: A Sketch of Guang'an Xiaoping Cadre College, Sichuan Province)

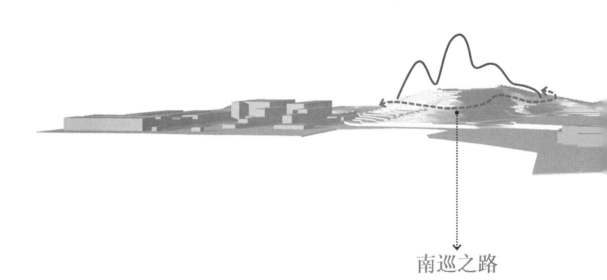

南巡之路

规划以山为骨骼水为脉络，最大限度地保留自然山体，依山就势布局规划，用一种与环境融合的低姿态，表达对场地的尊重和对邓小平朴实无华的呼应，将改革开放这个"春天的故事"融入景观设计中去，打造一个充满希望和未来的春天故事园。带领参与者学习伟人精神，重走伟人之路，拓展视野，打造开放校园，将周边区域整体考虑，以水生动，融会贯通。

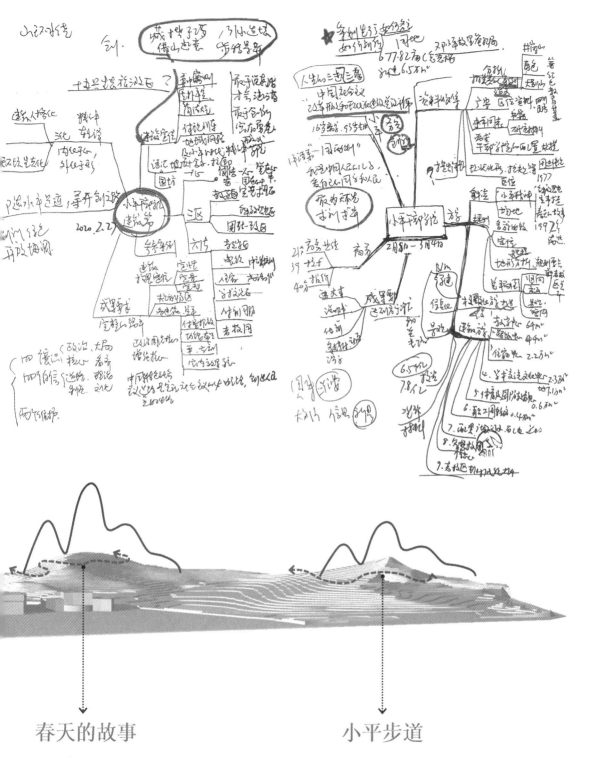

春天的故事　　　小平步道

The plan takes mountains as bones and water as the context, preserves the natural mountains to the greatest extent, and makes the layout according to the mountain with a low profile that agrees with the environment, expressing respect for the site and echoing the unpretentiousness of Comrade Xiaoping, The "Story in the Spring" of the reform and opening-up is integrated into the landscape design to create a spring garden full of hope and future. We lead participants to learn the spirit of Deng Xiaoping as a great man, retake his road, expand his horizons, create an open campus, consider the surrounding area as a whole, and bring vitality with water, perceive how and why.

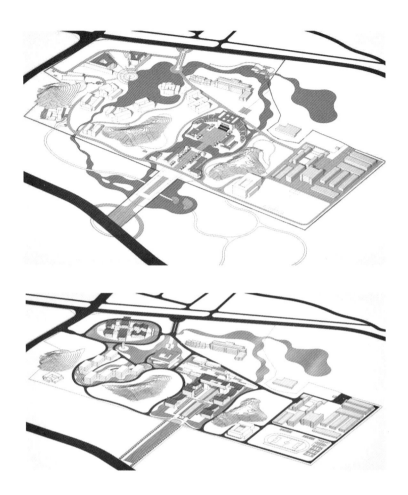

方案一设计理念：依山就势，族群消隐

方案整体规划依山就势，建筑与自然和谐共生。场地以党性教育的庄严对称作为学院西入口的设计手法，并采用地景上升的建筑设计用以表现邓公脚踏实地、扎根土壤的务实精神。学术文化交流中心融入了邓公开拓创新、改革开放的创新精神，打造山环水绕的自然生态学院。

方案二设计理念：开拓创新、同质异构

设计方案顺应地形和轴线，采用对称式布局与自由式布局相结合的方式灵活布置，各个建筑群体通过自然山水进行串联联系。

中部教学中心区结合地形特点，利用西侧长条地块，采取三进制院落布局，具有强烈的秩序感和仪式感，空间较庄重沉稳。

方案三设计理念：山环水绕，中轴对称以改革开放为设计宗旨，用水体软分隔场地，打造绿色、生态的开放式校园。建筑结合地形追求灵活性与功能实用性。使每栋建筑都有着自己的语言与独特形象。

The design concept of Scheme 1: Follow the Mountain Trend

The overall plan is based on the trend of the mountain, and the architecture is in perfect harmony with the nature. The site takes the solemnity and symmetry of Party spirit education as the design technique for the west entrance to the college, and the architectural design of rising landscape is adopted to show Deng Xiaoping's pragmatic spirit. The Academic and Cultural Exchange Center has integrated Deng Xiaoping's innovative spirit of pioneering and innovation, reform and opening up, and built a school of natural ecology surrounded by mountains and waters.

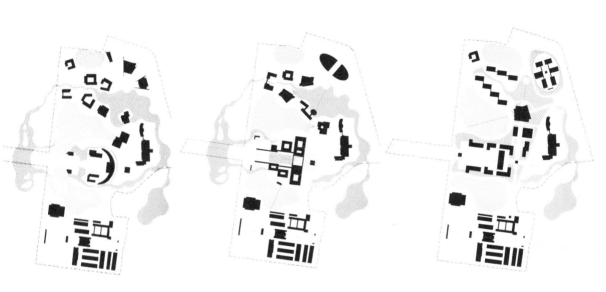

The design concept of Scheme 2: Explore and Innovate, the Same Quality but Different Structures

The design agrees with the landform and the axis. The symmetrical layout and freestyle layout are combined and flexibly arranged with each building group connected through the natural landscape. In line with the terrain characteristics, the central teaching center uses the west-strip plot and adopts the three-courtyard layout, which has a strong sense of order and rites, and the space is fairly solemn and tranquil.

The design concept of Scheme 3: Surrounded by Waters and Mountains with Central-Axis Symmetry

With the reform and opening up as the design purpose, we divide the site with waters to create a green and ecological open campus. Architecture combines terrain to pursue flexibility and functional practicality so that each building has its own language and a unique image.

学院整体设计采用"解放思想，展翅飞翔"的设计思路，秉承"传承经典，突破创新"的表现手法，以"虚实相生，刚柔并济"的表达方式呈现出来。

The overall design of the college adopts the design idea of "emancipate the mind, spread its wings and fly". Adhering to the expression method of "inheriting classics and seeking innovation", and present the mode of "virtual and real coexistence, and combination of rigidness and flexibility".

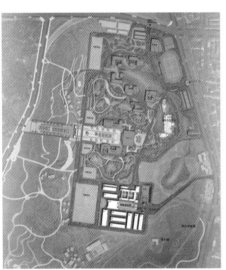

学院整体设计采用"春天的故事，改革开放"的设计思路，秉承"传承经典，突破创新"的表现手法，以"方圆交融，刚柔并济"的表达方式呈现出来。

The overall design of the college adopts the design idea of "Stories in spring , reform and opening-up. "Adhering to the expression method of "Inheriting classics, and seeking innovation", and present the mode of "Virtual and real coexistence, and combination of rigidness and flexibility".

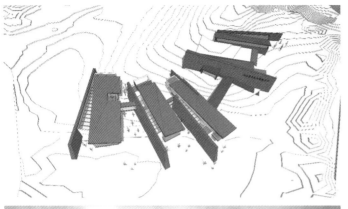

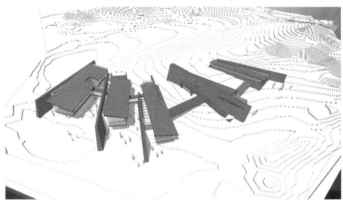

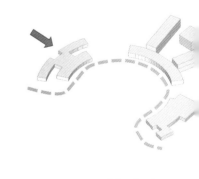

体块环绕围合

山石赋形，自然建筑

学术交流文化中心作为面向城市界面的形象展示，既要契合地域环境，又需要拥有独特的形象标示性。

建筑形态模拟山石，试图营造一种与自然共生破土而出的气势。建筑设计提取川东民居建筑元素，解构重组并用现代手法进行转译，通过坡屋顶的不同组合形态、庭院、架空空间以及与地形景观的融合来体现其建筑特征。

Mountain Stone Form-givins, Natural Architecture

As an urban landmark, the Academic Exchange Cultural Center should not only fit the local environment, but have a unique image mark. The architectural form simulates mountains and rocks, trying to create a ground-breaking momentum in symbiosis with Nature. The building borrows architectural elements from the eastern Sichuan residential buildings, deconstructs and reorganizes them, and translates them with modern techniques. It reflects its architectural characteristics through a different combinations of sloping roofs, the organization of courtyards, overhead spaces, and the integration with the terrain and landscape.

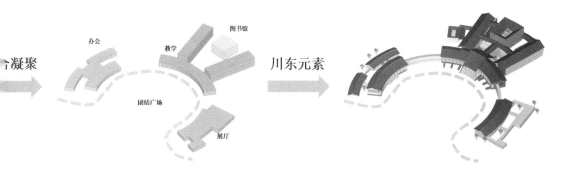

功能空间置入   传统元素置入

感知与包容
Perception and Tolerance

楚之晟 · 智慧运营中心
Chu Zhisheng Smart Operation Center

以"智慧·科技·创意"之名，激发一座城市的精彩，用互联网思维诠释艺术的价值、文化的价值与生活的价值，打造互动景观体验智慧运营中心。基地建设是利用新一代信息与通信技术来感知、监测、分析、控制、整合基地各关键环节的资源，在此基础上实现对各种需求做出智慧的响应，使整体的运行具备自我组织、自我运行、自我优化的能力，为基地创造一个绿色、和谐的发展环境，提供高效、便捷、个性化的发展空间。连续开放的多维度空间，它拥有开敞空间的呼吸感，尺度自由，步移景异的联动空间。云平台是"共享"与"自由"的精神运用于空间构成和服务体系。整个基地的低层部分都是连成一片的共享云空间，以此来呈现社会化、数字化和生态化的城市社区。

大雪

苍茫大地一片白，素夜田园生天籁。
潇潇铺飘雪满天，瑞兆来岁必丰年。

——当代·贵谷子《二十四节气之大雪》

楚之晟·智慧运营中心

Chu Zhisheng Smart Operation Center

用地面积：
20000m²
建筑面积：
59958.35m²
容积率：
2.5
设计时间：
2021

楚之晟智慧运营中心项目位于汨罗市，场地北临规划玉池路，南边为汨罗江大道，街对面有已建成的奥特莱斯商业广场，东侧是高泉北路，西侧为规划道路，交通十分便利。场地周边江面景观资源丰富，北望汨罗江，坐拥城市之景，江景、市景成环绕之势。项目用地呈不规则四边形，整个用地东西向最长处约200m，南北向最长处约为110m，均为荒地，场地内部较为平整。

该项目以"汨罗心、智慧谷"为形象定位，打造立体科技云城，连续开放的多维度空间，它拥有开敞空间的呼吸感，尺度自由，步移景异的联动空间。云平台是"共享"与"自由"的精神运用于空间构成和服务体系。整个基地的低层部分都是连成一片的共享的云空间。呈现社会化、数字化和生态化的城市社区。

（下图：楚之晟智慧运营中心总图）

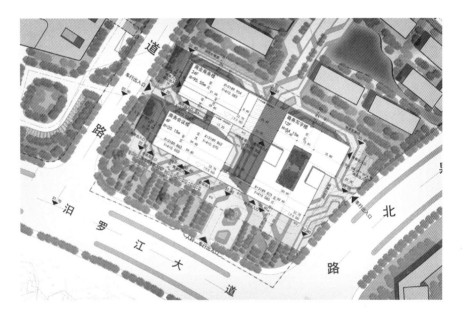

Site Area：
20,000m²
GFA：
59958.35m²
FAR：
2.5
Design Time：
2021

The project of Chuzhisheng Smart Operation Center is located in the city of Miluo, with the planned Yuchi Road in the north, Miluojiang Avenue in the south, the established outlet commercial plaza on the opposite side, Gaoquan North Road in the east, and the planned road in the west. Transportation is very convenient. Surrounding the site is abundant landscape resources. Looking into north to the Miluo River, it has a great view of the city. The river-city views are surrounded. The project land is in an irregular quadrilateral shape. The longest part of the whole land is about 200m from east to west, and the longest from north to south is about 110m. Currently, the land is wasted, and the interior of the site is relatively flat.

This project is positioned with the image of "Miluo Heart and Wisdom Valley" to create a cloud-city of three-dimensional technology, a continuous open multi-dimensional space, which has the breath of open space, free scale, and a linked space with different moving scenes. The cloud platform is the application of the spirit of "sharing" and "freedom" to space composition and service system. The lower part of the entire base is a connected shared cloud space. Present a socialized, digitized and ecological urban community.

(Below: A general plan of Chuzhisheng Smart Operation Center)

五大目标：
智慧生活·全民共享
城市治理·全网覆盖
政务协同·全市通办
生态宜居·服务连接
基础设施·提质融合

激活场地
（最大化场地视线和滨路、滨水优势，激活场地）
创建活跃的商务场所
（创建视线通廊，提升场地价值，创建一个活跃的商务场所）
创建综合接待中心
（融合酒店接待、休闲娱乐、商业购物、健身美食等多元功能为
一体，打造综合接待中心）
平衡解决方案
（转换场地尺寸小、进深小等限制条件，通过设计量身定制平衡
解决方案）

Five major objectives:

Smart living · Shared by all

Urban governance · Full network coverage

Government affairs coordination · City-wide access

Ecological livability · Service connection

Infrastructure · Quality enhancement and integration

Activate the site

(maximize the site-line of sight, roadside, and waterside advantages
and activate the site)

Create an active business place

(create a sight corridor, enhance the value of the site, and create an
active business place)

Create a comprehensive reception center

(integrating multiple functions such as hotel reception, leisure and
entertainment, commercial shopping, fitness and delicious food,
and build a comprehensive reception center)

Balanced solution

(change the limited conditions such as small site size and small
depth, and customize a balanced solution)

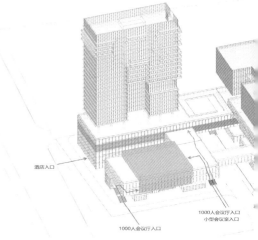

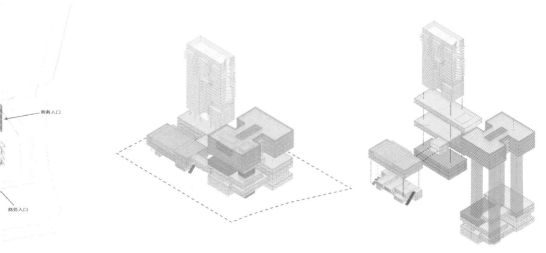

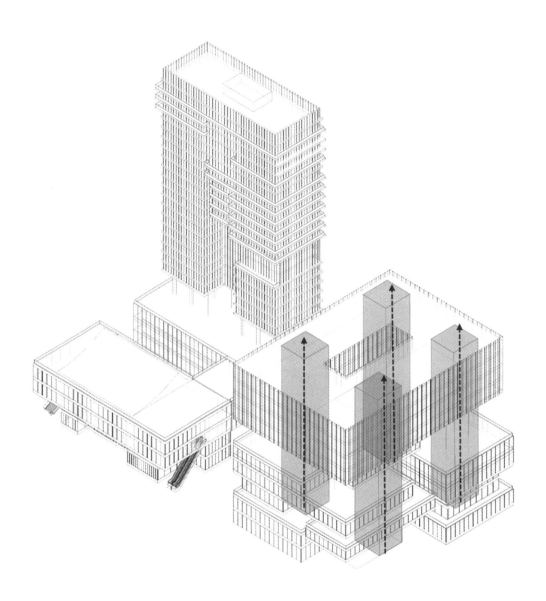

错落有致的建筑楼
优点：建筑体块丰富多变
　　　空间更具灵活性
　　　分部门使用核心筒，人行分流

Well-proportioned buildings with such
advantages:
varied building blocks
spatial flexibility
differentiated shaft application, and flow distribution

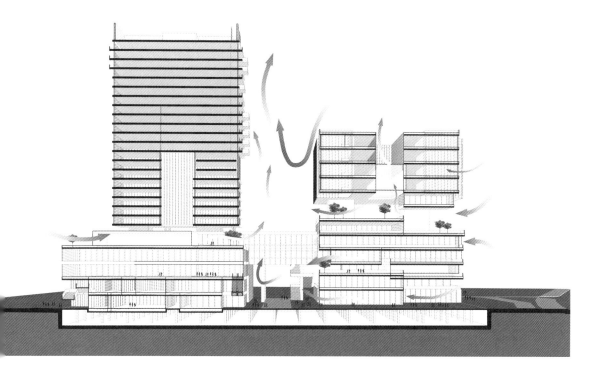

绿建设计:
提高围护结构热工性能，减少建筑能耗。
建造绿色屋顶，提高能源效率（夏季降温、冬季保温），增加视觉美观，辅助减少"城市热岛效应"。
采用双层通风玻璃幕墙，由通风层向室内输送新鲜空气，优化建筑通风质量。无须专用的机械设备，完全靠自然通风。
绿化墙体，消除噪声，提高周围空气质量，隔绝室外热源，降低室内温度，减少能量消耗及资金消耗。

green building design:
improve the thermal performance of enclosure structure and reduce building energy consumption.

build green roofs to improve energy efficiency (cooling in summer and heat preservation in winter), increase visual beauty and help reduce the "urban heat island effect".

the double-layer ventilated glass curtain wall is used to deliver fresh air from the ventilation layer to the room so as to optimize the ventilation quality of the building. There is no need for dedicated mechanical equipment and relies entirely on natural ventilation.

to cover the wall with green, eliminate noise, improve surrounding air quality, isolate outdoor heat sources, reduce indoor temperature, and reduce energy consumption and capital consumption.

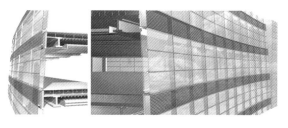

意义的载体
Carrier of meaning

株洲市信息港
Zhuzhou City Information Port

智慧园区建设是利用新一代信息与通信技术来感知、监测、分析、控制、整合园区各关键环节的资源，在此基础上实现对各种需求做出智慧的响应，使园区整体的运行具备自我组织、自我运行、自我优化的能力，为园区企业创造一个绿色、和谐的发展环境，提供高效、便捷、个性化的发展空间。我们的想法是通过"无忧畅行""森林换乘""株洲第一印象""24 小时活力环""生态步行网络""多孔的城市公共空间"等设计策略，通过一体化开发、功能混合打造，将株洲信息港打造成一个大型的立体城市公园、城市客厅。从"生态交通、水资源利用、生态能源、绿色建筑"的角度出发，将城市居民的健康、城市发展与自然环境的和谐共生作为设计目标，构建了与当地气候相适应、 能源高效利用、交通便捷、尊重自然的生态片区。

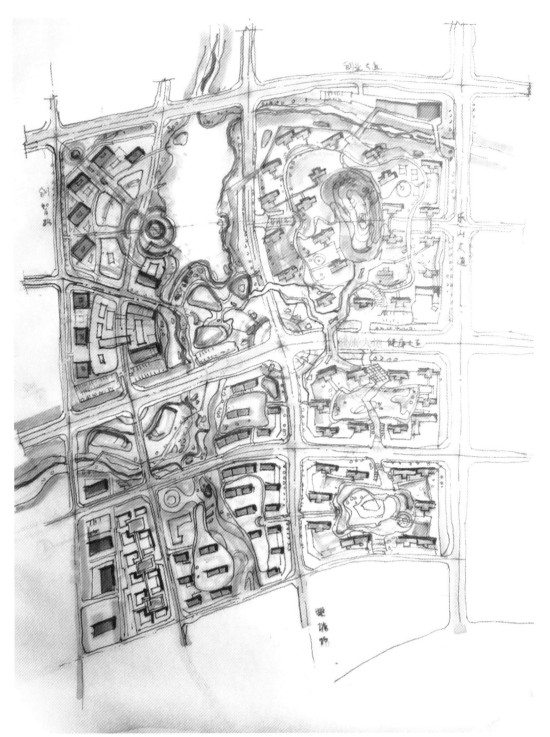

冬至

日照数九冬至天，清霜风高未辞岁。

又是一个平衡日，子线从南向北回。

——当代·贵谷子《二十四节气之冬至》

311

株洲市信息港

Zhuzhou City Information Port

用地面积：
497154.49m²
建筑面积：
1323086.69 m²
容 积 率：
2.67
设计时间：
2019

立体科技云城

云城是以促进"关联"为核心要旨，以"云服务"的方式给客户提供服务，是一种多维度的立体的科技之城。云平台是"共享"与"自由"的精神运用于空间构成和服务体系。整个产业园的低层部分都是连成一片的共享的云空间。

生态智慧之城

智慧园区建设是利用新一代信息与通信技术来感知、监测、分析、控制、整合园区各关键环节的资源，在此基础上实现对各种需求，并对需求做出智慧响应，使园区整体的运行具备自我组织、自我运行、自我优化的能力，为园区企业创造一个绿色、和谐的发展环境，提供高效、便捷、个性化的发展空间。

(下图：鸟瞰图)

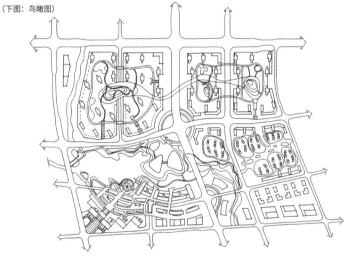

Site Area：
497,154.49m²
GFA：
1,323,086.69 m²
FAR：
2.67
Design Time：
2019

The cloud city of three-dimensional technology

The cloud city is a city of multi-dimensional solid technology based on the principle of promoting "association" and providing users with "cloud service". The cloud platform is the spirit of "sharing" and "freedom" applied to the space composition and service system. The lower part of the entire industrial park is a shared cloud space.The Smart city of Ecology

The smart park construction serves to use the new generation of information and communication technology to sense, monitor, analyze, control and integrate the resources of the key links of the park. On this basis, it can respond intelligently to various needs and make the whole park operate by itself. The ability to organize, self-run, and self-optimize creates a green and harmonious development environment for the park enterprises, providing efficient, convenient, and personalized development space.

(Below: Bird's eye view Terrain Model)

方案从"生态交通、水资源利用、生态能源、绿色建筑"的角度出发，将城市居民的健康、城市发展与自然环境的和谐共生作为设计目标，构建了与当地气候相适应、能源高效利用、交通便捷、尊重自然的生态片区。

利用新一代信息与通信技术来感知 、监测、分析 、控制、整合园区各关键环节的资源，打造一个生态智慧园区。

把"共享"与"自由"的精神运用于空间构成和服务体系中，使整个产业园的低层部分都是连成一片的共享的空间。呈现社会化 、数字化和生态化的城市社区。

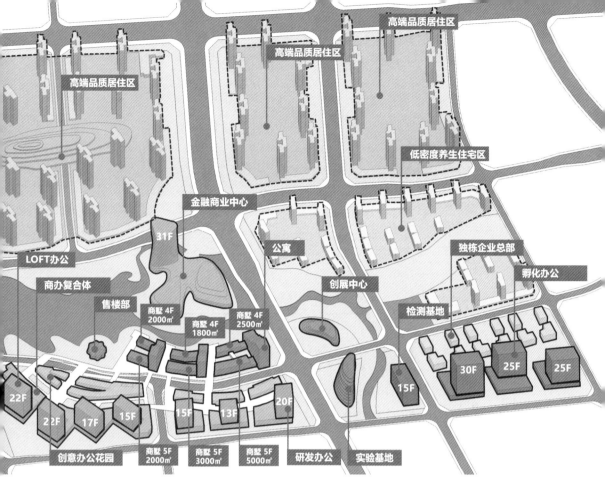

高端品质居住区

高端品质居住区

高端品质居住区

低密度养生住宅区

金融商业中心

31F

LOFT办公

公寓

独栋企业总部

商办复合体

创展中心

孵化办公

售楼部

检测基地

商墅 4F
2000㎡

商墅 4F
1800㎡

商墅 4F
2500㎡

22F

15F

30F

25F

25F

22F

17F

15F

15F

13F

20F

创意办公花园

商墅 5F
2000㎡

商墅 5F
3000㎡

商墅 5F
5000㎡

研发办公

实验基地

From the perspective of "ecological transportation, water resource utilization, ecological energy, and green buildings", the plan takes the health of urban residents, the harmonious coexistence of urban development and the natural environment as the designed goals and builds a structure which is compatible with the local climate, and uses energy efficiently. Ecological film with convenient transportation and respect for nature. Use the new generation of information and communication technology to sense, monitor, analyze, control and integrate resources of each key link in the park to create an ecological smart park.

The spirit of "sharing" and "freedom" is applied to the space composition and service system so that the lower part of the entire industrial park is a connected shared cloud space. Present socialized, digitized and ecological urban communities.

当下社会生活愈发多元化、复合化，设计采用复合型建筑来满足城市生活的社会性、动态性、多样性、复杂性，各构成部分优化配置于完整的系统中，并合理有序地运行。

At present, social life is becoming more and more diversified and complex. Composite buildings are designed to meet the sociability, dynamics, diversity, and complexity of urban life. Each component is optimally configured in a complete system, operating reasonably and orderly.

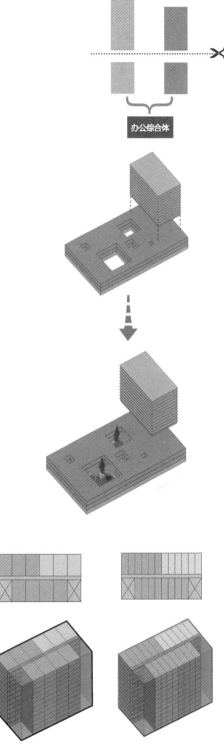

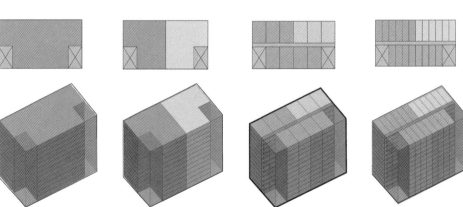

将共享绿色生态空间融入办公建筑内部，各层均有共享平台，外廊办公区可双面采光，外立面两种表情，整体立面和退台立面各不相同，使员工可以在绿色空间中工作。

The shared green ecological space is integrated into the interior of the office building, and each floor has a shared platform. The outer corridor office area can be day-lit on both sides, and the facade has two expressions. The overall facade and the retreated facade are different so that employees can work in a green space.

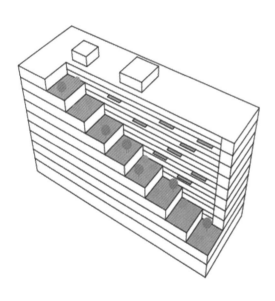

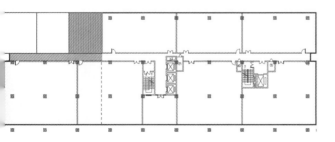

无穷与迭代

Infinity and iteration

西部（重庆）科学城科学会堂

West (Chongqing) Science City Science Hall

为区别传统会堂的设计手法，我们提出了未来"∞"（无穷大）的设计手法。寓意科学殿堂将用建筑连接现实场景，用科技连接未来的无限可能。以时空之舟的设计理念营造回溯亘古、畅游未来的意境，以科学桥梁的设计理念营造立足当下、探索未知的意境。我们把城市的自然空间、文化历史资产和科学科研成果提取为科学会堂的设计元素，并转化为科学会展活跃发展的机遇；为更好的发展高新区核心驱动力，我们设想以地域的"会展、文化、性格、生态"与科技的"神秘 X"相融合，创造出多元的科学会堂，打造未来的科学建筑；以创造一个有关科学的核心地标、一个与高新区相融合的科学殿堂、一个与环境友好的生态会堂、一个成渝城市群示范性场馆。

小寒料峭，一番春意换年芳。蛾儿雪柳风光。
开尽星桥铁锁，平地泻银潢。
——南北朝·王寂《望月婆罗门引·元夕作》

西部（重庆）科学城科学会堂

West (Chongqing) Science City Science Hall

用地面积：
246666.64m²
建筑面积：
334570m²
容积率：
1.07
设计时间：
2021

重庆科学城位于高新区直管园科学大道与高新大道交叉口西北角，距重庆西站约 10km，距江北机场 36km。基地距离银昆高速下道口约 700m，东临快速路科学大道、南接快速路高新大道，且南侧紧邻轨道站（规划 7 号线轨道站），交通便捷，可达性好。

重庆科学城是重庆市落实"一带一路"、长江经济带和"成渝经济圈"经济发展方针的具体举措，是实现"两点"定位、"两低两高"目标，实现科学发展、创新发展、生态发展的重要载体，是重庆未来发展新引擎。

（下图：西部（重庆）科学城科学会堂总图）

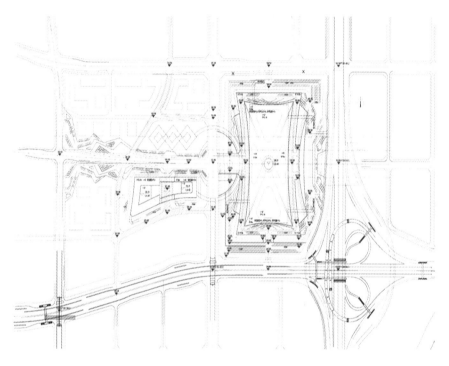

Site Area：
246,666.64m²
GFA：
334,570m²
FAR：
1.07
Design Time：
2021

Chongqing Science City is located at the northwest corner of the intersection of Science Avenue and Gaoxin Avenue in Zhiguan Park, High-tech Zone, about 10km away from Chongqing West Railway Station and 36km away from Jiangbei Airport. The base is about 700m away from the exit of Yin-Kun Expressway. It is adjacent to Expressway Science Avenue in the east, Expressway Gaoxin Avenue in the south, and adjacent to the rail station (Planned Line 7 rail station) to the south. The traffic is convenient and the accessibility is good.

Chongqing Science City is a concrete measure taken by Chongqing Municipality to implement the "Belt and Road Initiantive", Yangtze River Economic Belt, and "Chengdu-Chongqing Economic Circle" strategies. It is an important carrier for achieving scientific, innovative, and ecological development. It is also the new engine of future development in Chongqing.

(Below: General map of the Science Hall of the Western Chongqing Science City)

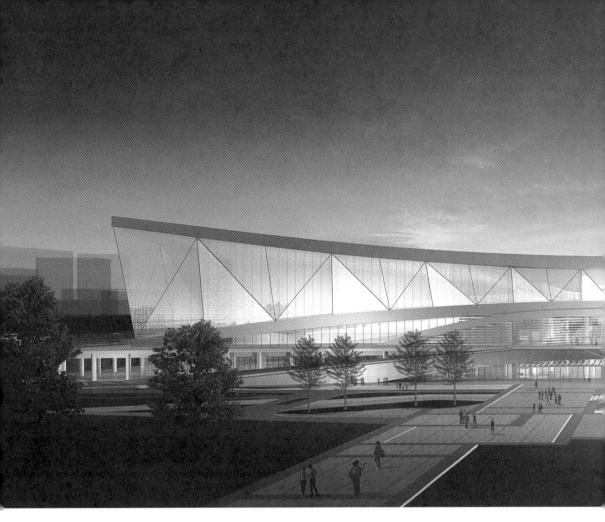

The design inspiration of the scheme comes from "bridge", which refers to the communication bridge that can promote the establishmen of cooperative relations between people and regions and spur up friendly exchanges. The design method of" ∞ "(infinity) in the future i proposed. It means that the palace of science is connected with real scene via architecture and with the future for infinite possibility vi technology.

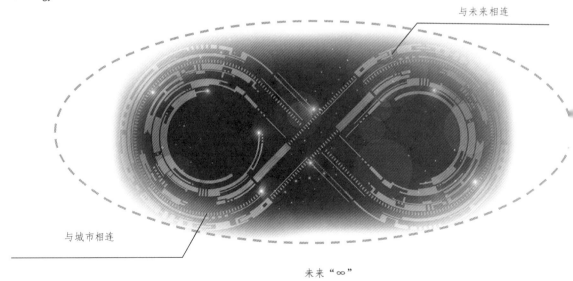

与未来相连

与城市相连

未来"∞"

…方案的设计灵感来源"桥
梁",可以促进人与人、
地区与地区之间建立合
作关系、带来友好交流
的沟通桥梁。提出了未
来"∞"（无穷大）的
设计手法。寓意科学殿
堂将用建筑连接现实场
景,用科技连接未来的
无限可能。

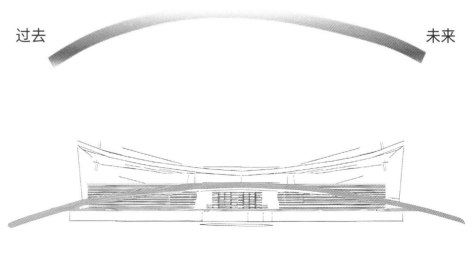

现在

过去　　　　　　　　　　　　　未来

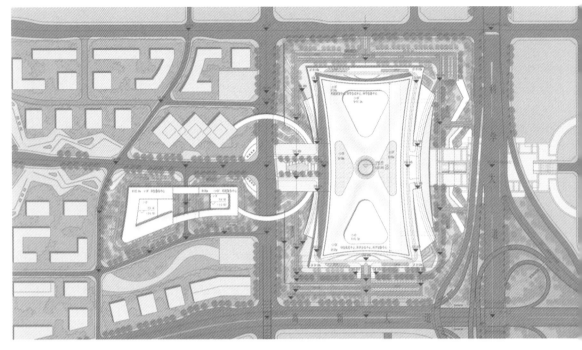

以山之形，融山之意

以桥之形，融桥之势

以水之形，融水之气

设计以"连接重庆的山与水，连接过去与科学未来的桥梁"为主题，通过生态仪式的景观体验，以及现代、简洁、流动的绿茵空间，打造一个具有仪式感、共享、生态的会议中心。设计元素以简洁的几何与线条形式呈现，将自然融入建筑，用体验创造场景。

The design is based on the theme of "connecting mountains and water of Chongqing, connecting the past and the scientific future", and the bridge through the landscape experience of ecological rites, and the modern, simple, and flowing green space, to create a ritual, shared, and ecological conference center. The design elements are presented in the form of simple geometry and lines, incorporating the nature into the building and creating scenes with experience.

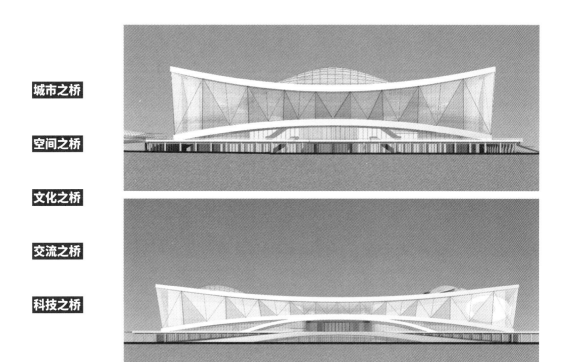

城市之桥

空间之桥

文化之桥

交流之桥

科技之桥

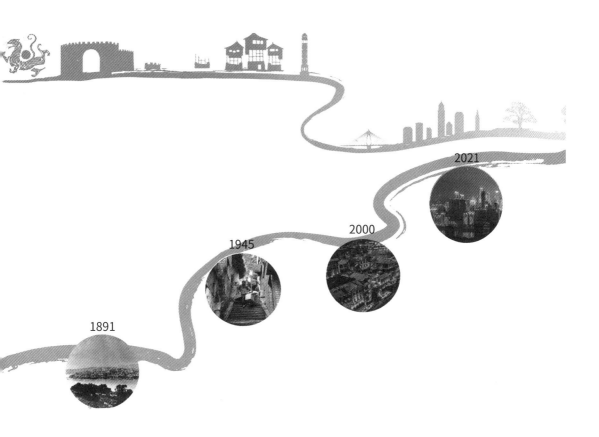

1891

1945

2000

2021

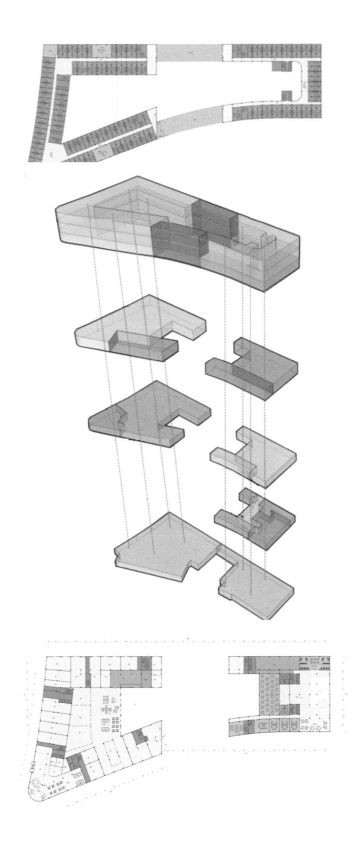

BIM 设计

根据原始地形图纸建立本项目三维地形模型，针对场地的高程、流域、汇水等情况进行数据分析，为项目的竖向设计提供参考依据。通过定义模型不同阶段的精度，规定模型每个阶段的表现特点，从而得到模型的可视化表达方案。在设计阶段，重点在于管线和空间查错、模型信息的外部表现、管线复杂区域、高大空间区域、交通空间等。

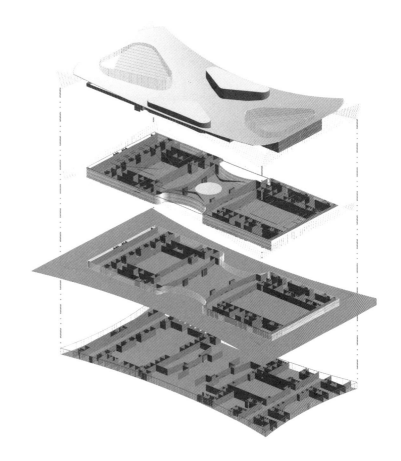

BIM design

Establish a three-dimensional terrain model of the project based on the original terrain drawings, and make data analysis of site elevation, watershed, catchment, etc., to provide a reference for the vertical design of the project. By defining the accuracy of different stages of the model and specifying the performance characteristics of each stage of the model, the visualized expression scheme of the model can be obtained. In the design stage, the focus is on pipeline and space error checking, the external performance of model information, complex pipeline areas, large space areas, and traffic spaces.

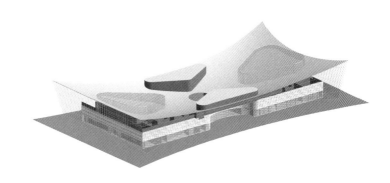

历史的回望
A look back in history

天下谷源展示中心
Tianxia Guyuan Exhibition Center

一颗稻谷的故事：稻谷驯化过程，见证着人类文明的发展过程，是文明的诞生，从原始生活→农耕时代→手工业，就像刚刚萌芽的幼苗，从泥土中来，带来希望，带来热情，带来勃勃生机。大米不仅仅是一种食物，更是人们幸福感的一部分。通过规划展示中心功能布局和游览流线，以稻谷、陶片为主题，深度结合玉蟾岩考古发现成果，突出"天下谷源·人间陶本"的主题意境。突出历史的传承与延续，对糖厂的老厂房"拆建结合·合理利用"对现状树木尽量保留。保留乡村自然肌理，外围以"原乡稻田·自然景观"打造农耕场景，再回首，回顾先民生活。

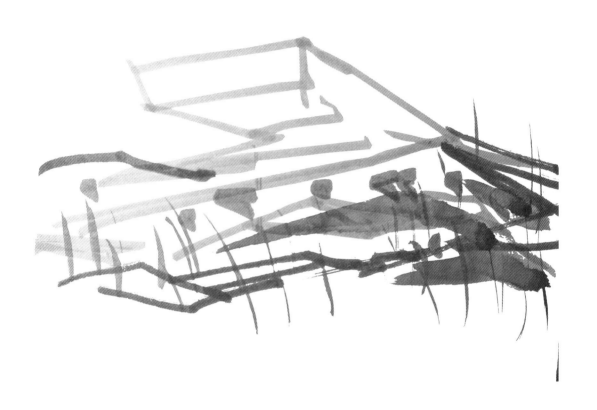

大寒

微湿易干沙软路，大寒却暖雪晴天。

——宋·陈著《游慈云》

永州道县·天下谷源展示中心

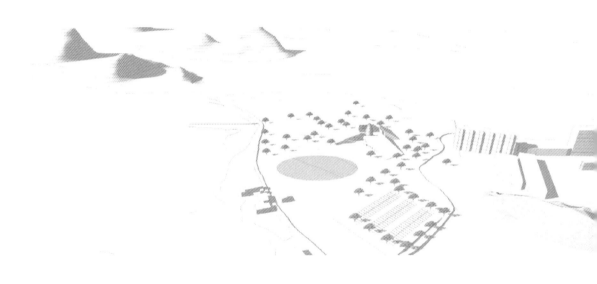

Yongzhou Dao County・Tianxia Guyuan Exhibition Center

用地面积：
107000m²
建筑面积：
22795m²
容积率：
0.21
设计时间：
2021

该项目位于永州市道县寿雁镇，项目选址用地 4800m²（72 亩）。西北侧以玉蟾岩为中心的保护区，约 47000m²（691 亩）。东南侧工贸中专用地总共 23150m²（347 亩）。拟规划总用地 107000m²（1605 亩）。通过对天下谷源展示中心的建设和文化展示片区的打造，引导寿雁镇整体规划格局的调整，使其成为寿雁小城镇的一部分，城镇格局渐趋明晰和合理；带动周边土地升值，打造寿雁小城镇建设的引爆点。

（下图：永州道县·天下谷源展示中心总图）

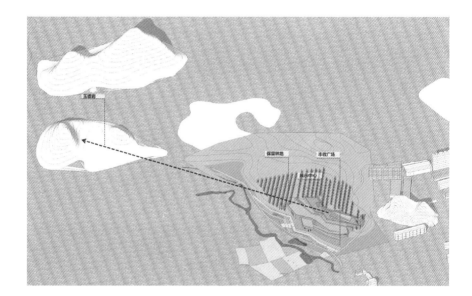

Site Area：
107,000m²
CFA：
22,795m²
FAR：
0.21
Design Time：
2021

The project is located in Shouyan Town, Dao County, the City of Yongzhou, with a site area of 4800m². The protected area centered on Yuchan Rock on the northwest side is about 47000m². A total of 23150m² of land is dedicated to industry and trade on the southeast side. The planned total land area is 107000m². With the construction of the Tianxia Guyuan Exhibition Center and the creation of the cultural exhibition area, the overall planning of Shouyan Town will be adjusted to make it a part of the small town of Shouyan, and the town pattern will gradually become clear and reasonable, hence promoting the value of the surrounding land and creating a tipping point of small town construction in Shouyan.

(Below: A General plan of Yongzhou Dao County• Tianxia Guyuan Exhibition Center)

"破土陶片"

建筑以陶皿破土而出的形态为灵感，立面以散落的水稻胚芽为基本元素，提取出基本的建筑类型元素并加以拓扑转换，形成形态丰富且独特的建筑空间。场地内尽可能保留树木，与周边的阡陌稻田景观交相辉映，营造出此处特有的场所感。

设计以夯土形态为建筑蓝本，以农耕文明及原始稻田景观为景观蓝本，是城市快速扩张的时代对农业景观的浪漫挽回和乡村故土的情感回归。

"Broken clay"

The building is inspired by the form of architectural pottery dishes having broken through the earth, and the facade utilizes scattered rice germs as the basic elements of basic architectural type plus topological transformation to form a rich and unique architectural space. Trees within the side shall be preserved as much as possible, complementing the landscape of the surrounding paddy fields and creating a unique place.

The design takes the form of rammed earth as the architectural blueprint, the agricultural civilization and the original rice field landscape as the landscape blueprint. It is the romantic recovery of agricultural landscape and the emotional return of rural hometowns in the era of rapid urbanization.

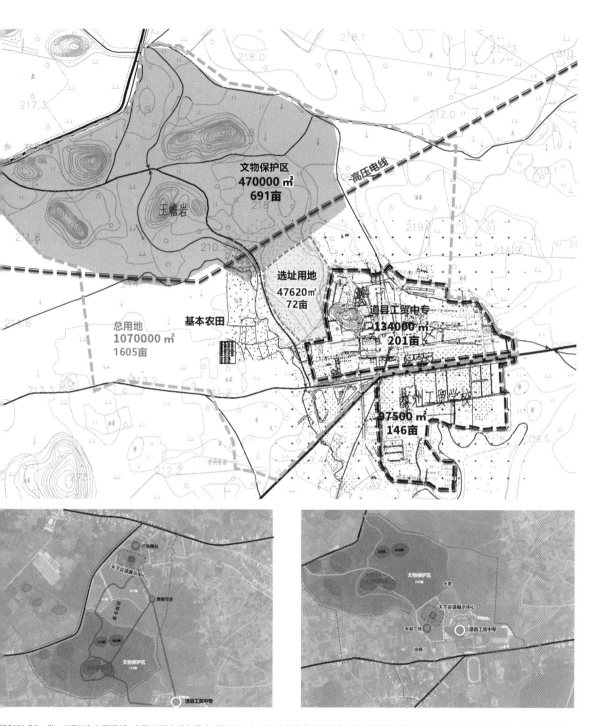

规划形成"一带一环"两条主要游线，串联公园内所有景点。展示中心与用于丰收节的广场舞台、西南方的稻田景观、东北方的水体景观形成一条景观中轴线，与西北侧的玉蟾岩遗址、东南方的道县工贸中专形成另一条次轴线。

is planned to form two main tourist routes of "the Belt and Road Initiative", connecting all the scenic spots in the park. The exhibition center, together with the square stage used for the harvest festival, the rice field landscape in the southwest, and the water landscape in the northeast, contribute to the central axis of landscapes and to another axis with Yuchanyan site in the northwest and the Industrial and Trade Secondary School of Dao County in the southeast.

方案一特点：
布局较为围合庭院保留现有树木，旧建筑通
过改造之后可与新建筑紧密结合，打开的两
个广场可举办不同活动。

方案二特点：
建筑造型有特色，分区明确，场所感强建筑
占地广。

方案三特点：
建筑整体性强，通过天井保留现有树木，北
侧结合稻田景观营造出景观式建筑，分区明
确，建筑造型有特色，轴线感强烈，与景观
结合紧密。

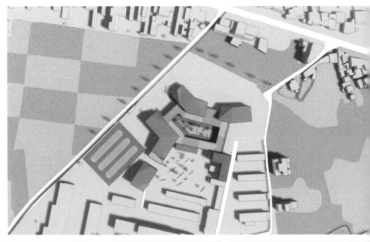

Features of Scheme I:
The layout is relatively enclosed, the
courtyard retains the existing trees, and the
old buildings can be closely combined with
the new buildings after transformation,
These two open squares can be used for
different activities.

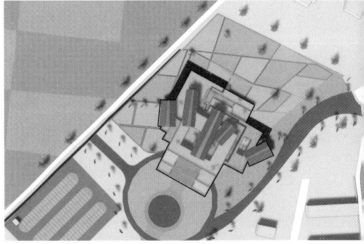

Features of Scheme II:
The architectural style is distinctive, the
division is clear, the sense of place is strong,
and the building covers a wide area.

Features of Scheme III:
The building is marked by strong integrity.
The existing trees are retained through the
patio, and the northern side is combined with
the rice field landscape to create a landscape
architecture with clear division, distinctive
architectural modeling, a strong sense of axis,
and close combination with the landscape.

339

Combine

《尔雅》：妃、合、会，对也

《说文解字》：合口也。从亼从口。候合切

变幻的景象

不变的初心

抱扑归真

设计是一场心灵的修行

对话与合作

激励与尝试

不负时光，珍惜信任

唤醒心中的山水城市

寻找到自己的生活原点

Find the origin of your life - An interview with Fanyi

原文刊载于"湖南勘察设计"

学——是自我完善与超越的过程

持续不断地学习，不是为了拿更多的证书来证明自己，而是希望不断完善自己的知识结构，将理论与实践相互对照，一方面寻找解答工作中存在的困惑，另一方面寻找到自己的价值表达与设计感悟。学习是一种自我完善与不断超越的过程，学习也不仅是单向的输入，更是需要分享交流输出，才能达到更好的学习效果。

通过搭建学习型团队，经常性地组织读书会、优秀设计及演讲比赛，与年轻工程师们分享学习经历和项目经验。帮助、带动和培养一批优秀的年轻设计师，帮助他们成长，不断完善自我，超越自我。人的一生非常短暂，想学的又太多，希望不断有新的蜕变。

行——设计是一场持续的修行

规划、建筑、景观、室内设计等不应该被人为地划定为不同专业性范畴，作为一名合格的建筑师，从宏观的城市视角到微观的一草一木都可以被纳入设计，设计不应设限，建筑师可以跨界，设计更应该回归到自己的初心，建造出令人感动，给人温暖的幸福建筑。设计、阅读、旅行应是每年工作计划中的关键词，建筑师阅读的深度与广度，影响着建筑师的工作方法及价值取向，阅读更是长年累月养成的一种习惯。

像海绵般吸收的目的在于实践，行胜于言，这一理念也印证在多年主持设计和负责的项目中。在大量项目的历练过程中，从相对单纯的建筑设计领域，跨度到总体规划、城市设计，并延伸到周边环境地块的景观设计，逐步包括室内装饰设计等，通过对项目全方位的整体把控和设计意图的完美实现，完成了大量的优秀项目，也成为省内为数不多的凭借优质设计能力得到业界认可的设计团队，特别是在做石门市民之家项目时，我们意图将石门秀美山水元素融入建筑设计中，使建筑仿佛与大地景观融合，形成诗意天成、浪漫自由的状态。

建筑师的角色就像电影导演，需要全面调度精心策划才可能产生一部巨作，也像音乐会的指挥家，必须纵横捭阖胸有成竹才能形成有灵魂的乐章作品，总建筑师在不同项目中需要扮演不同的重要角色，每个项目都需要长期的经验积累、设计灵感的捕捉、匠人追求完美的精神以及团队的通力合作才能完美展现；但每个项目永远也是留有或多或少的缺憾，这时候往往需要总建筑师有高度自省的能力，既要打破惯性思维，也需要对过往经验和成绩的反思、批判、重新审视。

建筑设计是一个非常注重团队合作的行业，任何一个成功设计背后的荣誉都是无数的汗水和心血凝聚而成的。感恩成长的道路离不开身边优秀的老师和团队成员，正是他们的激励和信任，不断鼓舞其前行，需要专注一心、吐故纳新、持续精进。其实，前面的工作过程需要一切清零回归原点，设计之路只是刚刚开始。

创——工作需要整体性思维及不断创新

当 2006 年 AutoCAD Revit 1.0 刚开始在省内推行，就觉得技术创新在未来的设计行业中存在着巨大潜力，马上着手在单位组织学习和培训。市场经过了十年摸索，BIM 建筑信息化模型真正开始在设计中发挥积极作用，不仅努力探索在负责项目中 BIM 的应用，同时也积极推动在建科院建立起企业级运用的品牌与典范，2015 年参加第十四届北京住博会 BIM 大赛，我们的武陵山文化产业园项目获得一等奖，2016 湘西州武陵山文化产业园项目（非物质文化遗产展览综合大楼、数码影视文化艺术中心）荣获湖南省首届 BIM 技术应用大赛二等奖。

在各方不断地推动下，院里组建了 BIM 中心，派人员去中国建筑设计研究院学习 C-BIM 体系，而且在 2016 年与长沙高新区规划局、长沙规划信息中心、东方红建筑公司等数十家单位组建了湖南建科 BIM 产业链发展联盟。推动 BIM 的发展不仅仅是设计单位的事情，更应该从全产业链角度发挥巨大价值，设计与施工，专业分工及行业割裂严重，需要 BIM 这样的技术来整合，建筑师应具备整体性思维设计作为前端从项目前期就在线开始掌控，可以更好地保证大型项目的完成及后期运维效果。随着 BIM 产业链产业发展联盟的推进工作，2016 年又启动了建科院与省内各大高校的产、学、研、平台搭建工作，希望建科院在建筑行业人才培养，科技成果转化方向有所建树，与湖南大学、湖南工业大学、湖南工学院都建立了有关 BIM 与住宅产业化等多个课程的战略合作计划。

建筑行业的知识可谓是学无止境，只有抱着终生学习，不断探索的信念，才能脚踏实地，不断创新，才有可能成为优秀的既懂设计又懂管理的复合型人才。如何创造一个和谐的、生态的人居环境是每一个对建筑设计有责任感的设计师的不懈追求，这也正是未来不断努力的方向！

"寻找自己的初心，寻找自己的原点，寻找我们最初对这个城市的感受"

Find our original intention, origin and initial feelings about the city

城市与建筑师的原点——访湖南省建筑科学研究院总建筑师周湘华 [J]. 高青，周湘华 . 中外建筑 . 2017(03)

| 凡 益 | 每一位建筑师心中都有一座理想的城市，这座城市可能是现实存在的，也可能是自己的"理想国"，那么理想的城市离不开优美、雅致、趣味的街区，而这些街区一定是有自己故事的，那也就是我们所说的历史街区，接下来能否请您先聊下历史街区对城市发展的影响呢？ |

| 周湘华 | 我在长沙参加过大大小小很多个项目的评审，特别是一些历史街区方面的项目评审，也参加过长沙市规划局组织的多次针对长沙市历史文化特色要如何体现的专家咨询和讨论会议。如何体现长沙特色这也是我们本土设计师最纠结的一个地方，到底长沙的特色是什么？我们经常说长沙是山水洲城，但是现代长沙的城市特色几乎是没有的，特别是经过文夕大火之后，长沙的历史文化脉络被毁，历史文化街区逐渐消逝。当然长沙现在也有一些历史街区在建设，如太平街、都正街，也取得了不错的效果，但是比起上海、杭州、成都等这些城市，我们的历史文化街区的建设还有很多要提高的地方。 |

| 凡 益 | 您觉得有哪些当前长沙这座城市的历史街区有些地方需要提升？可以考虑采取哪些措施？ |

| 周湘华 | 就目前而言，我觉得有 3 个比较重要的点。第一点是开发和保护的平衡，既要考虑城市的文化和历史的保护，也要进行城市的开发和更新；第二点是要延续，也就是历史文脉的延续和相关政策的延续性，每一届政府都会对城市开发提出新的亮点和变化， |

但是我希望对历史文化街区的保护或者改造的是一个完整连续的过程；第三点是要激活，历史文化街区不是一个在博物馆的古董，是要跟现代生活相结合的，通过注入新的商业模式和运营模式，提升整个历史街区的活力。

凡　益　　　　　　　　　当前国内许多青年建筑师通过建筑绘本的形式来表现城市趣味的多元性，您如何看待这一行为？

周湘华　　我觉得这是非常有创意的想法。我去年看过北京几个类似的项目，就非常有意思。这种通过这种漫画、绘本的方式能让更多大众来更好的认识和了解我们的城市。现在很多人对自己城市的了解其实并不多的。对于像我们这些看着漫画成长的 70 后们，对于绘本有着很深的情结，通过绘本的形式来表现城市、规划和建筑，是非常有意义的事情，而且以后也可以衍伸到其他形式的产品，比如文创产品或者游戏开发，让更多的人来使用和了解我们所表达的东西。

凡　益　　　　　　　　　每个建筑师都可以像导演一样去演绎一部巨作；也可以像音乐家一样指挥各种乐器产生不同的音符；更可以像厨师一样去烹饪一道道美食，所以建筑师都是在创作幸福的感觉，如果您变化角色来诠释这三种职业，您如何诠释？

周湘华　　影：建筑师的角色就像电影导演，从最开始剧本的选择推敲到演员角色的筛选以及后期的剪辑，场景的组合，背景音乐的气氛烘托，一道道、一层层，反复修改，多人参与，最终呈现在观众眼前的才能是让人欢喜、让人忧回味无穷的经典梦幻。

声：建筑师也像指挥家，轻重缓急、抑扬顿挫、千回百转，各种声部、乐器、演奏者都需要相互协调配合，当一个演出团队从最开始的杂乱无章到训练有序，声音从无法控制到优美和谐，需要耐心也更需要无止境的重复千百回。

味：建筑师也像厨师，酸、甜、苦、辣、咸，五味杂陈，建设方、施工方、监理方、材料商，你方唱罢我登场，各自调配着不同的味道，建筑师不仅品尝着自己作品生成过程中的各种滋味，更是被"调味"同时也是"被设计"的对象。

凡　益　　　　　　　　　作为省级大院的总建筑师，相信您的作品数量和质量都是非常高的，能否请您介绍一下自己的创作心得？比如最近做的一个项目？

周湘华　　这个项目我们表达的是对湘西地形地貌特质的诠释。我们把博物馆设计成山石形状，就是想把历史文化用一个坚固的"容器"来装载，从而像磐石一样稳固厚重。

凡　益　　　　　　　　　这个项目对您来说与以往的项目有哪些不同？

周湘华　　这个项目的造型相对来说比较复杂，所以我们表达的建筑意象可能会跟其他建筑不一样，同时在建造方面我们也采用了 BIM 技术，对建筑的体型和空间结构进行分析，将室内、室外、展成以及今后的运营通过 BIM 总体综合在一起，希望建筑的完成度较高，最终能够把我们的设计思想贯彻进去。

凡　益　　　　　　　2013 年三联生活出版一本汉宝德的《如何欣赏建筑》，这本书诠释了大概 20 座让他感动的建筑，有名的包括泰姬陵、米拉公寓、悉尼歌剧院等，从艺术美学的角度审视建筑，发掘建筑的美应该是理性而感性的，是否能请您谈谈建筑与艺术之间联系？

周湘华　我觉得建筑就是艺术中间的一个组成部分吧，也是艺术非常重要的一个表达形式，建筑通过各种手段把艺术固化。此外，艺术可以给我们建筑师很多启示，很多艺术创作的天马行空，能够给建筑带来一些新的思路，通过艺术和生活的结合，使建筑能够更好地符合人们对于当下的需求，我希望今后的建筑能够成为一件艺术作品，能够让人看到建筑的光彩和亮点。

凡　益　　　　　　　当今建筑与艺术的互动是越来越多，城市双年展似乎是近来建筑与艺术最火的展览。2015 年深港双年展将以原始的"采集和狩猎"形式，而不是现代的规划和抽象化，来展示建筑和城市，您是否关注了这届深港双年展，有什么看法呢？对哪个部分印象比较深？

周湘华　深港双年展给我印象最深的是活动的开展和组织。展览的场地选在一个老面粉厂，艺术家、建筑师和一些文化创作组织在一起，不仅仅只是在做一个展览，还有很多的活动。比如说我参加的蛇口议事，他们会发一个绘本类的册子，以互联网的思维来增加公众的参与度，有一个展厅里面有很多的盒子，里面有很多的标签，可以供游客写下感受，同时会有一些专家采用圆桌议事的方式进行一个论坛。

凡　益　　　　　　　这届双年展将搜罗可以利用的、来自世界各地的实际项目和片段，建立社会和经济联系，从而打开环绕在我们周围的封闭结构，并提出如何利用现状基础设施共建未来，特别强调了公众的参与度。当中有没有您觉得很有启发性的项目？

周湘华　我觉得深港双年展的集装箱这个项目给我印象比较深刻，各企业的展馆都利用废弃集装箱做成的展馆，这不仅很好的契合了深圳这个港口城市的城市特色，而且也呼应了城市原点的展览主题，体现了再利用、再回收、再循环的概念。我们今后的一些设计理念可能也能从中吸取一些灵感吧。

凡　益　　　　　　　"城市原点"从策展官方的诠释是为如何建设，怎样建设城市，城市为谁而设？它的什么怎样的过去？现状如何？未来何去何从？您如何理解城市原点这个概念呢？

| 周湘华 | 我认为城市原点的概念就是回归吧。他们所说的城市原点实际上就是在寻找自己的初心，寻找自己的原点，寻找我们最初对这个城市的感受。 |

| 凡 益 | 城市的现状的再利用、再思考和再想象，并开放现有建筑，具体怎么来理解三种空间之间与城市原点的关系？ |

| 周湘华 | 我去参观展览，他们会通过多媒体和绘本的形式表达一个人在建筑里面的生活状态。再想到我们做建筑设计时，出发点是什么，为什么要做这个设计？这就是我们需要找到和回归的原点，可能以后更多的建筑师会意识到这些吧。 |

| 凡 益 | 那您觉得当今城市建设和建筑创作中应该回归的原点是什么？ |

| 周湘华 | 其实在这些年的城市建设中，我们实际上没有考虑多少绿色建筑和生态循环。其实城市最需要的就是绿色食品、新鲜的空气、良好的居住环境，这些就是我们要考虑的原点，我希望能够通过把这一个个的点连接成一个网络，把我们的城市建设的更美好。 |

| 凡 益 | 渐渐发现许多的建筑设计院开始转型，不论是技术层面还是人事层面，甚至业务多元化层面上都逐渐开始蔓延这种趋势，您所在的建科院正在转型吗？ |

| 周湘华 | 这确实是一个转型和升级的过程，今后的发展可能会突破传统设计院的模式，从而更加突出建筑师自身的特点和作用，而设计院可能会形成产品导向型的、技术导向型的、生产导向型的、产业导向型的和客户导向型的，像扎哈这些明星设计师就会专门去做一些特别有意思的作品，大部分的设计院就做生产型的工作，而对我们建科院来讲应该是属于技术型的，面对整个行业的形式，我们的想法是希望把我们院里面的科研和生产结合来进行相互推动。目前项目没有那么急了，我们就有更多的时间来对技术进行突破和创新，设计师也都有更多的时间静下心来进行思考，是不是能够有新技术、新发展方向的探索，通过这样一个双螺旋驱动方式来做，通过我们的科研来形成生产力，再将设计和生产结合在一起来推动发展。 |

| 凡 益 | 最后请您讲一下对青年建筑师的寄语。 |

| 周湘华 | 在未来，设计和艺术、设计和生活以及设计的跨界结合会越来越多。也许现在确实是建筑行业的冬天，既然冬天到了，那么春天也就不远了，我希望能尽快迎接建筑行业的春天，也希望现在的年轻人能够投身到现在的变革中间来。 |

使生活也成为艺术——《城市·环 境·设计》(UED) 杂志社
主办的常德老西门建筑品谈

Make life also art - "Urban Environment Design" (UED)Changde Laoximen Buildings sponsored

老西门的重生 [J]. 崔愷，孔宇航，魏春雨，刘伯英，齐欣，李保峰，朱剑飞，张鹏举，彭礼孝，石磊，周湘华，陈钊，曲雷，何勍，郑楠，李向北，贾蓉，柳青，方际三，钟晓曦. 城市环境设计 . 2016(04)

非常高兴今天能来参加这个研讨会，前面聆听了很多老师对这个项目观点和看法，作为湖南本土建筑师，也有幸参与了很多的历史街区改造项目，以及历史名人故居复原项目，同时也在做历史文化街区与新媒及文创很多相关方面的研究。我想在全国各地历史街区的保护与复兴都面临着很多同样的问题和困惑，许许多多的历史街区采用仿唐宋、仿明清的舞台布景道具做假古董的改造方式，缺乏对现代生活场景的再造。今天在这个研讨会的过程中我找到了一些解答，同时也让我有了更多思考。首先进入街区时，就被葫芦口水街上方造型特殊的人行桥惊艳到了，之后窨子屋中古今搭配、现代与传统的结合也让人感受到时空的穿越，回迁楼的屋顶花园与精心保护的旧木板上的老报纸，也使人体会到了设计师对当地居民人性化的关怀与匠心独运。虽然葫芦口广场景观人工化的痕迹略显过重，但作为醉月楼的观赏室外露天剧场，其设计还是有可圈可点之处，钵子茶馆红红的外墙带点粗野主义的建筑语汇，恰当地体现了湖南人特有的热情与豪放。大千井巷自由组合的屋顶与墙面也表现了湘西北地区人民粗犷与自由的特性。

这个项目很多方面可以供我们借鉴，譬如如何处理历史信息的真实性与创新性的矛盾，以及渐进式的改善当地居民基本生活与可持续的生存发展的要求。建筑师不仅仅要有基本的立场与价值，更要以开放与弹性的态度处理好建筑细节与场所再造的关系。今天何总讲述了从感性方面如何在项目中注入情感以求打动业主，并且通过建筑师在现场参与式设计将设计想法贯穿始终。我认为我们可能需要先对项目有较为深入的理解再积极与业主沟通可实现方式，最终项目落地性才强。第二点我认为目前来说城市的改造具有许多新的模式。像刚刚崔总提到的整个项目呈现出一种"碎片化"设计的形式,这种"碎片化"如何凝练出一个整体的面貌,可能也是建筑师需要思考的。老西门改造项目以窨子屋为起点,再拓展到其他的街区巷道,所以我认为把碎片化的单独项目如何有效串联起来形成整体也是重要的一点。

第三点我认为是运营的问题,我们接触到的很多项目设计做得很成功,但是后续运营并不一定成功,往往原因就在于没有在最初就把运营单位引入项目中,导致后期运营困难,门可罗雀。老西门这个项目在最开始就兼顾了运营的问题,完善最后理念的实施和落地环节,成功激发了社区的活力。这也说明作为城市复兴项目应该是需要一个全盘的思考和策划。

最后有几点建议,是否在该片区的更新过程中能尽量利用质量还不错,有保护改造价值的原有建筑,而不是全部拆除新建,适当保持该片区的居民的历史记忆,历史的传承应该是生活记忆的传承,真实生活场景的延续,单独追求传统元素符号化的片段复原缺乏生命活力。其次应将非物质文化遗产的动态参与式的展示与商业活动相结合。如举办定时定点的街头传统戏曲表演将文化活动与运营相结合并通过网络新媒体进行推广。

建议内部的水系应结合立体绿化、雨水收集,老护城河等。内部水体的循环自净与水生植物的生物净化方式等生态技术,将老西门项目做成常德旧城改造与海绵城市相结合的示范点。并让更多的市民了解与参与更多的历史街区的改造工作,才能最终实现城市的复兴。

谢谢!

凡益开题 |BIM 的未来在哪里？

Where is the future of BIM?

凡益开题第九期:BIM 的未来在哪里 ?[J]. 中外建筑 . 2017,(11)

周湘华，湖南省建筑科学研究院副院长、总建筑师。

张平，湖南省建筑科学研究院 BIM 技术应用中心主任。

姚守俨，中国建筑第八工程局有限公司 BIM 工作站站长，教授级高级工程师。

田华，中国建筑第五工程局有限公司 BIM 中心主任。

彭子茂，湖南省交通职业技术学院建筑工程学院副教授。

张 彬，上海德晟建筑工程公司任公司董事长兼总经理。

周湘华

1975 年，"BIM 之父"——佐治亚理工大学的 Charles Eastman 教授创建了 BIM 理念至今，BIM 技术的研究经历了三大阶段：萌芽阶段、产生阶段和发展阶段。BIM 技术可以实现建筑工程的可视化和量化分析，提高工程建设效率。在国内，一些大型的设计企业、施工企业早在 10 年 8 年前就开始关注和研究 BIM 技术的应用。到今天 BIM 技术广为人知，越来越多的企业成立 BIM 部门研究和应用 BIM 技术，住房和城乡建设部和地方政府陆续出台了 BIM 的相关政策文件，BIM 已经成为行业发展的趋势。住房和城乡建设部副部长易军说："谁掌握 BIM，谁就拥有未来"。那么 BIM 的未来在哪里呢？

应该来说，很多企业和项目使用 BIM 技术取得了良好效果。地产公司以万达为代，成立了自己的 BIM 部门，并创造性研发万达 BIM 总发包管理平台，革命性推行万达 BIM 总发包管理模式；设计院以中国建筑设计研究院为代表，研发了 CBIM 设计系统；施工单位以中建为代表，也进行了施工企业 BIM 平台的研发。可以说这些企业使用 BIM 技术都取得了一定的成果，企业从技术到管理上都得到了很大的提升。从整个行业越来越多的 BIM 大赛来看，参赛作品越来越多，质量越来越高，都说明了 BIM 技术的应用取得了可喜的成效。是不是说 BIM 技术在行业中已经成熟应用了呢？抛开表面的热度，进行冷静的思考，我们也会发现 BIM 发挥真正价值还存在瓶颈，距离真正落地还有很长的路要走。你会发现：很多 BIM 大赛的参赛材料千篇一律；很多项目应用都还是基础的、碎片化的 BIM 应用，很多 BIM 运维的产品系统还不够成熟；很多中小企业推广 BIM 举步维艰；BIM 技术在设计施工运维方面应用有待于进一步加强；基层的人员没有使用 BIM 的动力；传统流程和新技术新管理产生的冲突。不同企业面临不同问题，关注点也不尽相同。作为设计企业来讲，用 BIM 做设计，应该是用 BIM 做三维设计而不是做二维设计，BIM 的核心是模型数据为主，二维图形为辅。但设计院交付的成果是必须满足国家二维制图规范的图纸，如果不改变现状，减少二维的工作，不从根本上普及 BIM 三维设计，设计院在现阶段很难普及 BIM 设计。设计企业的 BIM 之路究竟应该如何继续走下去？

作为施工企业来说，BIM 可视化的虚拟建造的确为工程建设提前规避掉了很多问题，减少了很多不必要的浪费，直观的三维模型也为项目各参与方之间沟通、项目方案的优化提供了很多便利。但是，施工企业 BIM 应用的价值远不止于此。施工企业 BIM 的未来应该如何继续进行？

BIM 是一种技术，更是管理和流程的再造，用老一套思维使用 BIM 没有动力，容易导致失败。如何破局，探索 BIM 技术在设计、施工乃至运维的未来之路，实现 BIM 的历史使命，推动设计与建造的技术进步，推动行业信息化的历史进程。

| 张 平 | 自 2002 年 BIM 作为概念引入中国，最开始就是在设计行业开始推广的，至今已有 15 个年头。虽然有很多专家和设计企业宣称已经实现了"BIM 的正向设计""基于 BIM 打通了整个行业链"、甚至"BIM 的建筑全生命周期应用"等目标，但事实上，BIM 的应用仍然停留在可视化与碰撞冲突检测等有限层面上。实现全专业协同的 BIM 正向设计案例寥寥无几，更不用说基于 BIM 技术的设计企业信息化建设成果了。为什么经过了近 15 年的发展，设计院还不能普及 BIM 技术开展设计工作呢？这里面有着各种方面的原因。其中，非常重要的一点原因就是设计院普遍没有真正下功夫将 BIM 理念应用于项目开展和设计管理中去，致使 BIM 技术应用在这么多年过去之后依旧只停留在表象层面。明朝圣人阳明先生曾说："破山中贼易，破心中贼难"。如何真正从实现 BIM 价值角度出发推动设计院 BIM 技术普及是一个非常值得思考的课题。在此，我有以下几点拙见。

第一、必须调动起设计人员的主动性。人，是技术应用和普及的主体。必须让设计人员切实了解和体会到 BIM 对于设计工作是具有重大价值的，才能使得设计师们真正主动地来学 BIM，来用 BIM，来研究 BIM。

第二、推动建立以三维模型为核心的出图标准和规范，以及相关的报建审批机制。应用 BIM 开展设计，自然是做三维设计。但是设计院交付的成果是"图纸"，图纸必须符合国家的二维制图规范和标准。如果基于三维的设计要求交付二维设计的成果，那设计师根本没有动力去先做 BIM 设计再出二维图纸。所以，必须建立"以三维模型为核心的出图标准和规范"，让二维图纸直接从模型中"切出来"；实现"三维模型"和"二维图纸"的一致性关联，摒弃掉不必要的二维标注化表达规范和标准。这样，才是给设计师们"减负"，才真正有利于推动 BIM 设计。相关的报建审批机制建立，也是同样的道理。

第三、通过更深、更精细化、更优化的设计，配合政府指导建议，提高设计取费。BIM 技术在设计院发展缓慢有一个很现实的原因就是：传统的设计手段已经能满足我绝大部分的设计需求，那为什么还需要去学习一个新的、陌生的东西来增加自己的负担呢？采用 BIM 技术能为我们带来更高质量的设计成果，但是，如果更高质量的设计成果不能带能更高的企业收益，企业自然没有动力去做这种"费力不赚钱"的事。所以，更高品质的设计产品应该换来更高的设计收费，企业才有动力下真功夫普及 BIM 设计工作。

破山中贼易，破心中贼难。只有真正落地地推行 BIM 技术在项目和企业中的应用，不断深入挖掘 BIM 的价值，才能真正实现 BIM 的高度普及和企业效益的大大提升。 |
| 姚守俨 | "按模施工"即按照 BIM 模型完成工程建造。包括施工策划、施工组织、竣工交付等施工全过程。"按模施工"是施工生产未来的必然选择。 |

首先，现行的"按图施工"是"按模施工"的特例。施工图是利用平、立、剖方式表达物体的空间关系和联系，需要多张施工图按一定规则才能表达准确。而 BIM 的三维模型优越于二维施工图表达形式，其可视化、模拟性等降低了业主的识别难度（所见即所得）。

其次，从建筑物表达方式演变来看。中国古代建筑施工基本是师徒之间口传身授，虽然《营造法式》也有建造图，但是主要集中在中式且定式的建筑物。而西式建筑的传入，带来了现代施工图，为个性化建筑物识别提供公认的方法。当二维施工图方式无法表达建筑物特殊、异性结构时，人们创造出 2.5 维施工图来弥补正视图的表达缺陷。但是，2.5 维图纸无法展示建筑物内部空间关系。因此，三维 BIM 模型表达方式不仅克服了上述缺点，而且保证了数据的唯一。

最后，BIM+ 技术发展引起施工方式变革。例如：BIM+ 增强现实技术。增强现实支持将设计投射回到现实世界，以便了解其将给环境造成的实际影响。实景模拟施工主要是在前一道工序完成情况的基础上进行下一道工序虚拟施工模拟，通过模拟分析，得出各工序之间的逻辑关系，从而确定合理的施工方案来指导施工。因此，这种虚实结合的施工方式，有利于提高施工效率和准确性。

"按模施工"有助于施工行业的工业化生产组织施工生产模式，比较接近工业生方式产中的柔性生产，柔性生产包括虚拟生产和拟实生产。拟实生产即"拟实产品开发，它运用仿真、建模、虚拟现实等技术，提供三维可视环境，从产品设计思想的产生、设计、研发、到生产制造全过程进行模拟，以实现在实体产品生产制造以前，就能准确预估产品功能及生产工艺性，掌握产品实现方法，减少产品的投入、降低产品开发及生产制造成本"。这与 BIM 在建筑全生命期的作用十分贴合，"按模制造"在制造业生产中早已实现，所以"按模施工"有助于施工行业工业化。

BIM 是建筑构件的核心数据，是打造以数据驱动为核心的关键信息。其中，BIM 构件的尺寸、坐标等空间数据，通过接口导入数字化、智能化的生产设备、加工设备、施工设备中，最终完成建筑部品的生产，促进 BIM 技术与工业技术深度融合入。同时，BIM 是施工组织的基础数据，可以为产品分解、任务分解、工期计划、材料管理、设备管理、成本管理、过程控制等提供相关信息。实现各参与方信息共享，转变过去粗放的模式，走向信息化、精细化、工业化。

科技在飞速发展，科技在改变我们的生产方式，基于"按模施工"的新式施工组织模式将引领建筑信息技术迈向更坚实且更辉煌的未来！

田 华 　　近几年，BIM 技术得到大力推广和应用，风靡整个建筑行业。从住房和城乡建设部到地方政府纷纷给予高度关注，并出台各类指导意见，强力推行基于 BIM 的建筑信息化的应用和普及。为响应"号召，"顺应"潮流"，各大建筑企业纷纷派人外出学习、调研。一时之间，各类关于 BIM 应用的咨询公司应运而生，各类关于 BIM 的论坛和会议也层出不穷，整个建筑行业呈现出一片争先恐后学 BIM 用 BIM 的繁荣景象。然而，沉下心来冷静客观思考，会发现在很多施工企业，BIM 技术有点食之无味弃之可惜的感觉。

　　很多项目都在变着花样的使用 BIM 技术，4D 不够高大上，就上 5D、6D。越来越多的项目注重高端的应用和漂亮的报告，却放掉了一个最基础的东西——模型质量。这个基础的不稳定，直接导致 BIM 应用成了中看不中用的花架子。想要 BIM 流程能够真正实施起来，这些流程必须有一个实施的基础，那就是模型。模型的质量很重要，没有一个优质的模型作为基础的话，那无论流程设计的多好，执行的标准有多严格，都没法做出有实际意义的事情来。对于施工企业而言，模型的正确性更是体现在 BIM 模型与实际施工的一致性，许多模型中常常出现数据错误，模型与现场实际情况不符，或是设计修改之后没有更新模型等情况，这样的模型对于实际施工起不到应有的帮助。

　　导致模型的质量不高的因素有两个，一是作为依据的设计图纸有错漏，二是建模人员的能力和对图纸的理解不够导致的二次错误。怎么才能保证模型的质量呢？那就是必须进行深化设计。深化设计是指施工单位在建设单位提供的合同图纸（模型）、技术要求和指定采用的相关标准、规范的基础上，结合材料、设备的实际尺寸和自身的施工工艺，综合协调各专业、各分承包商或设备供应商的技术需求，对其进行细化、补充、优化和完善，形成各专业的现场施工图纸（模型）或工厂加工图纸，同时对各专业设计图纸（模型）进行集成、协调、优化与校核，以满足现场施工及管理需要，编制必要的设计文件，送审、讨论、定稿、出版并经严格的发放程序分发相关文件的全过程。通过基于 BIM 的深化设计过程与其他承包商、分包商、设备供应商提前进行协调，取得彼此间的配合，可以弥补设计单位施工经验不足和对建筑材料市场了解不足导致的图纸错漏，弥补建模人员能力和理解不够导致的二次错误，有利于优化、完善建筑工程各系统的设计，提前解决现场施工有可能出现的问题。施工企业只有通过深化设计对模型进行完善，使得模型数据的正确性得到保证，对 BIM 模型中的各类信息加以充分利用才能得以实现，才会有进度、物料、工厂预制、工程造价等应用。所以，高质量的模型是后续施工中各种 BIM 应用的基础，只有以深化设计为抓手，切实提高模型质量才能促进 BIM 应用真正落地。

彭子茂 　　随着互联网技术的快速发展，BIM 技术在建筑业与其他领域的逐渐深入，建筑类企业和科研院所的发展急需掌握 BIM 技术的专业人才，而高校作为人才培养的第一阵营，传播 BIM 理念，培养更多适应行业需要的 BIM 技术复合型人才，提升学生就业核心竞争力，显得尤为重要。

一、高校 BIM 发展现状

随着 BIM 人才需求量的不断增加，各高校实施 BIM 高端人才培养的改革已经不可避免。BIM 贯穿于工程项目的整个寿命周期，BIM 技术涵盖了项目决策、设计、招标投标、造价、施工等各个工作领域的信息，这就需要 BIM 专业应用人才具有很强的工程设计、施工、管理的综合素质，甚至具有较为丰富的现场管理经验，对培养 BIM 专业应用人才的教学人才提出非常高的理论、实践能力要求。湖南省大多数本科和职业院校已经逐步开设 BIM 技术相关课程，但是 BIM 专业教学已滞后于行业发展，师资匮乏，课程编排滞后，BIM 教材真空，BIM 实训室建设不尽完善。因此，进行专业培养目标修订，合理设置课程标准，完善校企合作机制，加强师资队伍建设，是高校紧接行业、提升专业核心竞争力、增强社会服务功能的首要问题。

二、高校 BIM 技术人才培养方面的对策建议

（一）完善培养目标

BIM 技能型人才的培养，应在理论与实践相结合的基础上进行专业知识的传授，注重培养学生的工程实践能力，其培养目标的设置应顺应社会对 BIM 技术人才的需求，依托建筑工程专业的优势地位，进一步明确定位，不断寻求新的生长点，以 BIM 技术为协作平台，将课程内容和实践训练以任务形式嵌入到项目全寿命周期管理中，实现"做中学"，真正按照工程项目运行方式实现对学生 BIM 建模、工程策划、工程招标投标管理、工程造价文件编制、工程项目质量、安全和进度管理及项目运营等能力的训练和提高。

（二）调整课程设置

改变原有的传统专业课程教学结构，将 BIM 技术融入建筑工程课程教学，主要有两种途径：一是将 BIM 纳入一门或两门课程中；二是在几门课程的所有部分都涉及 BIM 技术。

第一种方法是单独开设一门或两门新的专业基础课，主要向学生介绍 BIM 软件使用中的基本概念。这种方法可能会使学生仅关注软件的操作使用，从而忽视 BIM 在项目整个生命周期中产生的作用。

第二种方法有利于学生形成系统、清晰的专业知识体系。可采用"以项目全寿命周期管理"为主线的课程体系，该课程体系包括项目全寿命周期的三个阶段：决策阶段、实施阶段及运营阶段。决策阶段：工程图学 III、建筑制图与识图、建设工程经济、建筑结构。BIM 建筑结构建模软件—Revit、BIM 机电模型软件—Magicad。实施阶段：工程估价、工程项目招投标与合同管理、建设工程法规；土木工程测量、土木工程施工、工程项目管理、工程造价管理、BIM 图形算量软件、

BIM 钢筋算量软件、BIM 计价软件、BIM 现场布置软件—GCB 、BIM5D；运营阶段：工程 BIM 模型综合应用、物业管理、管理学基础。在安排教学计划时可以将 BIM 技术引入到专业课程中，教师可以借助 BIM 技术中的结构设计和 VR 虚拟现实技术帮助学生理解专业知识。

（三）校企共建实践教学平台

根据不同的专业，开设与建筑工程技术专业职业岗位相应的实训教学模块，建立独立的教学标准与考核制度。建立技能实训平台，整合完善原有实训资源，实现组合优化、资源共享，突出真实的职场氛围，创设真实的工作环境。与合作企业共建校内外专项技能实训基地，让学生在仿真的职场环境和实际施工现场进行实训，培养学生从事各专业所必备的专项技能。与企业共同制定顶岗实习课程标准、教学计划、过程监控办法及考核标准。通过直接在施工现场进行的顶岗实习，让学生完成实际工作过程，从而形成职业综合技能。实践教学过程融入"虚拟施工"，构建基于施工全过程的 BIM 教学模型。

（四）师资队伍和 BIM 生产团队建设

通过实施"人员互聘、项目互助、能力互提"的师资队伍建设模式，大力提高教师的双师素质和"三能"水平（能进行职业教育教学设计；能指导学生实训、实践；参与企业相关研发工作，能帮助企业克服技术难题）。派遣校内现有专业带头人去国外或者企业现场学习实践，使他们及时跟踪建筑行业发展趋势和动态，准确把握建筑工程技术专业群建设与教学改革方向。同时，建立由行业专家、企业一线技术骨干组成的兼职教师资源库，参与到课程体系建设、课程教学和教材的编写等工作中来。

三、结语

高校应该抓住行业改革的机遇，结合自身特色将 BIM 技术有效引入到新的教学体系中，创建高质量的实训平台，加大师资队伍投入，提升教学质量，从硬件和软件两个方面提升院校的培养水平，只有如此才能够适应社会和行业的发展，才能为社会培养出高质量的 BIM 专业技术人才。

张 彬　　现在从事建筑工程行业的人员一提起 BIM，几乎无人不知，无人不晓，可 BIM 技术在互联网，大数据时代下，进展的越来越快，雾里看花导致了 BIM 万能论，无用论等一系列的认知。俗话说得好，外行看热闹，内行看门道，谨以此文阐述我对 BIM 技术发展的一些浅显认知。

BIM 首先是一个工具，三维的设计工具。在实际项目实施的过程中逐渐感觉 BIM 是一个建筑信息化的平台，是一个技术信息化平台。既然是信息化平台，怎么使用？

现在包括施工企业在内的很多企业开始使用 ERP 管理系统，一段时间后把它放在一边，没有让它产生最大效益，其中最关键的问题就是基础数据信息录入的缺失。很多企业 ERP，最后都没有能应用起来，一个原因缺少基础数据信息录取。另外就是企业执行力的问题。BIM 就是一个建筑行业的数据信息平台，未来要突破的就是怎么将这两个平台融合在一起，BIM 的基础信息平台能在 ERP 系统中能发挥更大的作用。所以说我们要打通技术平台和管理平台之间的瓶颈，让 BIM 在企业内部产生更大的效益。BIM 不仅仅是一个技术手段，还可以在管理上产生效益，比如说成本管理、计划管理、劳动力配置管理，包括质量、安全，都可以在各个环节中产生它的作用，这就是要一个企业各个部门动起来，未来在企业内部，不仅仅是技术部门在用，BIM 在管理中也要产生效益。

大数据的时代，有了 BIM 以后，从对模型的关注到信息的关注，再把 BIM 技术作为一个综合的系统，和其他相应的技术结合起来，我们发现在 BIM 平台之上产生了大量的数据，这些数据可以是结构化的，也可以是非结构化的。在这个阶段，无论是从我们交付服务的方式，还是未来客户提出需求的内容，或我们自己的生产方式和交付的产品都发生了根本性的变化。从数据库到数据仓库，我们遇到了前所未有的爆炸式的数字化的信息，而这些海量信息如何去运用，或者该用什么方式去运用？又如何将这些数据运用到我们工程建设行业的全生命周期？对整个工程建设行业而言，都需要重新来思考。一般而言，大数据具有庞大的数据规模、快速的数据流转、多样的数据类型、极富价值以及真实性等特征，是重要的信息资产，我们施工行业的大数据时代，实际上也是刚刚开始，它的数据积累还不充分，只是先提出一个理念，是一个未来发展的趋势。在目前阶段，首先是宣传一个理念，未来大数据能够带来什么。但现实是什么呢？建筑行业大数据的积累是远远不够的，还是一个基础建设阶段和推广阶段，很多人还没有接受推广的理念。一些一流的企业已经开始设立信息部来接受这个理念，逐渐把数据开始沉淀下来，但数据如何沉淀、如何归类，这些工作还有待于加强，但至少它已经有了这个理念，已经在做基本准备工作，这就是对先进技术的敏感性。还有很多的企业，对它们而言还处在推广阶段，所以说整个行业里面的发展水平是参差不齐的。

BIM 技术管理理念、建筑产业化装配式等技术的深入应用将提升建筑性能和质量，进而使居住和生活品质更上一个台阶。无论 BIM 将来如何发展，方向又会如何调整，我们会一直在 BIM 技术应用发展的道路上努力前行。

作品获奖
Works won

作品 1: 湘江风光带
　　　　1995 年荣获湖南省优秀设计一等奖（参与）

作品 2: 湖南省潇湘园 EPC 项目
　　　　1999 年在中国 99 昆明世界园艺博览会中获得突出贡献奖

作品 3: 湘西武陵山文化产业园博物馆项目
　　　　荣获 2015 年第十四届中国住博会最佳 BIM 设计应用一等奖

作品 4: 湘西州数码影视文化艺术中心
　　　　荣获 2016 年湖南省第一届 BIM 技术应用大赛设计组工程单项类二等奖
　　　　荣获 2017 年中国建设工程 BIM 大赛单项三等奖
　　　　荣获 2017 年第六届"龙图杯"全国 BIM 大赛设计组三等奖

作品 5: 湘西武陵山文化产业园二期（非物质文化遗产展览综合大楼）项目
　　　　荣获 2016 年湖南省第一届 BIM 技术应用大赛设计组工程单项类二等奖
　　　　荣获湖南省 2018 年度优秀工程设计二等奖

作品 6: 装配式建筑施工的 BIM 技术应用（课题）
　　　　荣获 2016 年湖南省第一届 BIM 技术应用大赛施工组单项类二等奖

作品 7: 湘西武陵山文化产业园城市设计
　　　　荣获 2017 年湖南省优秀城乡规划设计二等奖

作品 8: 石门县市民之家片区建设
　　　　荣获 2018 年第四届"科创杯"中国 BIM 技术交流暨优秀案例作品展示会大赛最佳 BIM 设计应用奖一等奖
　　　　荣获 2018 年第九届"创新杯"建筑信息模型（BIM）应用大赛文化旅游类 BIM 应用第三名
　　　　荣获 2018 年"龙图杯"第七届全国 BIM 大赛设计组三等奖
　　　　荣获 2021 年度湖南省优秀工程勘察设计建筑设计项目一等奖

作品 9: 腾讯双创社区（重庆高新）项目
　　　　荣获 2019 年"龙图杯"第八届全国 BIM 大赛设计组一等奖
　　　　荣获 2019 年第十届"创新杯"建筑信息模型（BIM）应用大赛科研办公类 BIM 应用第三名
　　　　荣获 2019 年第四届建设工程 BIM 大赛三类成果奖

作品 10: 湖南创意设计总部大厦
　　　　荣获 2019 年碧桂园杯首届湖南绿色建筑设计竞赛职业组方案类金奖
　　　　荣获 2020 年"龙图杯"第九届全国 BIM 大赛设计组二等奖
　　　　荣获 2020 年工程勘察设计质量管理小组活动成果大赛 I 类优胜成果

作品 11: 桂阳县全民健身中心场馆（A、B、C 区）
　　荣获 2019 年湖南省优秀工程勘察设计优秀建筑工程设计二等奖

作品 12: 城市智慧运营中心
　　荣获 2020 年碧桂园博意设计杯第二届湖南绿色建筑设计竞赛职业组方案类银奖

作品 13: 隆子县扎日乡曲桑边境小康村建设项目可行性研究报告
　　荣获 2020 年湖南省优秀工程咨询成果二等奖

作品 14: 隆子县玉麦幸福 - 美丽边境小康示范乡 EPC 建设项目
　　荣获 2021 年度湖南省优秀工程勘察设计工程总承包项目一等奖

作品 15: 山南市隆子县玉麦乡玉麦小康村建设项目
　　荣获 2021 年度湖南省优秀工程勘察设计建筑设计项目二等奖

作品 16: 隐宿
　　荣获 2021 年湖南省优秀农村住宅方案设计暨优秀民宿方案设计大奖赛民宿类一等奖

 01

 02

 03

2010

 01

文泰新城
岳阳市湘阴县
73823.98m²

Wentai New Town
Xiangyin County, Yueyang
73823.98m²

 02

福苑国际大酒店
张家界
66174.55m²

Fuyuan International Hotel
Zhangjiajie
66174.55m²

 03

湘建大厦
长沙长沙县
39594m²

Xiangjian Building
Changsha County, Changsha
39594m²

 A

汨罗市西片区控制性详细规划
汨罗市
5250800 m²

Controlled Detailed Planning of
the West Area of Miluo City
Miluo
5,250,800 m²

 B

湘阴县文化体育活动中心
岳阳市湘阴县
24150.8m²

Xiangyin County, Yueyang
Xiangyin County
24150.8m²

 C

东湖广场
浏阳市
360000m²

East Lake Plaza
Liuyang
360000m²

 D

望城一中图书馆
长沙市
4248.05m²

Wangcheng No. 1 Mi
School Library
Changsha
4248.05m²

2014

 A

 B

 C

 D

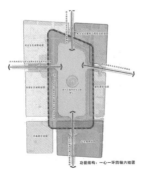

历年作品
Works over the years

参展
Exhibition

2015 年 12 月，蹊径与大道 ——中国新生代建筑师展览（长沙站巡展）开幕论坛，长沙

2016 年 12 月，大设计 新生态—省建筑师学会年会，长沙

2018 年 9 月，（第十届）全国建筑科研院所协作网会议，西安

2020 年 9 月，湖南省张家界市城市建设与 BIM 技术应用论坛及技术交流咨询会，张家界

2021 年 11 月，武汉设计日暨武汉设计双年展，武汉

2021 年 11 月，艺术之眼·后湖星潮 2021 长沙后湖艺术周

2021 年 12 月，中国科协农村专业技术服务中心及中国科协宣传文化部评选为年度科技志愿者先进典型

2021 年 12 月，湖南省科学技术学会授予 2021 年"科创中国"湖南省企业"创新达人"荣誉称号。

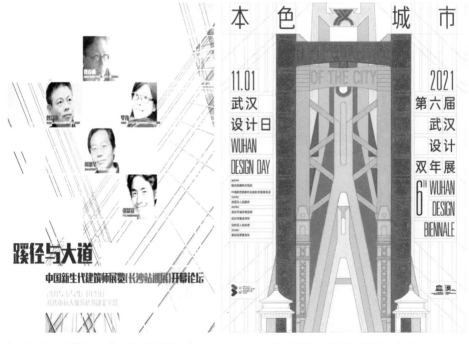

中国新生代建筑师展览（长沙站巡展）　　武汉设计日暨武汉设计双年展

行业影响力
Industry influence

2007 年 1 月，湖南省土木建筑学会　授予湖南省土木建筑学会第七届理事会理事

2008 年 12 月，长沙市人民政府办公厅　授予 2006~2008 年度长沙市社会主义新农村规划编制工作先进个人

2010 年 12 月，湖南省住房和城乡建设厅和湖南省人力资源和社会保障厅　授予湖南省优秀勘察设计师

2011 年 12 月，湖南省土木建筑学会　授予湖南省土木建筑学会第七届理事会先进个人

2012 年 3 月，湖南省土木建筑学会　授予湖南省土木建筑学会第八届理事会常务理事

2013 年 10 月，中国建筑学会　授予中国建筑学会资深会员

2014 年 12 月，湖南省建筑师学会　授予湖南省建筑师学会第四届理事会常务理事

2015 年 6 月，湖南省住房和城乡建设厅　授予湖南省新型城镇化标准设计图集编制评审专家库专家

2016 年 7 月，湖南工业大学　聘为湖南工业大学人居环境设计学专家硕士研究生指导老师

2017 年 10 月，湖南省国资委　授予湖南省国资委国有资本经营预算支出项目评审专家

2018 年 2 月，湖南建工集团有限公司　授予湖南建工集团 2017 年度"劳动模范"

2018 年 3 月，长沙市发展和改革委员会　授予长沙市发改委专家库专家

2018 年 7 月，湖南大学研究生院　聘为湖南大学专业学位硕士研究生校外指导教师

2018 年 12 月，湖南省人民政府　授予湖南省综合评标专家库评标专家

2018 年 12 月，湖南省城乡规划学会　授予改革开放四十年湖南省城乡规划行业发展贡献奖

2019 年 7 月，中共湖南建工集团有限公司委员会　授予湖南建工集团 2019 年国企楷模优秀个人

2020 年 9 月，中共湖南省委统战部　授予湖南省同心美丽乡村建设专家顾问团成员

2021 年 11 月，湖南省建设科技与建筑节能协会　授予 2021 年度湖南省建设行业"科技创新突出贡献人物"

2021 年 12 月，中国科协办公厅　授予中国科协 2021 年度科技志愿服务先进典型

2021 年 12 月，湖南省科学技术协会、湖南省科学技术厅，湖南省工业和信息化厅、湖南省工商业联合会 授予"科创中国"湖南省企业"创新达人"

2021 年 12 月，湖南省科学技术协会　授予 2021 年湖南省科协系统"最美科学传播者"

2022 年 1 月，长沙市勘察设计协会　授予 2021 年度长沙市勘察设计协会工作先进个人

研究成果
Research results

著作 1：2006 年 12 月《覆土型博物馆建筑研究》工程硕士论文
著作 2：2021 年 6 月《PDCA-BIM 设计全过程管理》中国建筑工业出版社
著作 3：2022 年 8 月《内观外筑》中国建筑工业出版社
著作 4：2022 年 12 月《绿色低碳设计与施工》朗文出版社

论文 1：《空间意蕴的构成》
　　　　2005 年 4 月《中外建筑》
论文 2：《苍梧之野，南风之时——舜帝陵园改扩建及总体规划设计》周湘华、陈立祝
　　　　2010 年 7 月《建筑知识》
论文 3：《从重引导到重实效——湖南省新农村住宅图集编制与回顾》
　　　　2010 年 7 月《城市建设》
论文 4：《"非常"安置—以滨湖农民安置小区为例》
　　　　2010 年 7 月《中外建筑》
论文 5：《中医医院改扩建更新策略》
　　　　2010 年 8 月《中外建筑》
论文 6：《周湘华作品精选》
　　　　2014 年 8 月《中外建筑》
论文 7：《文化产业园区地域特色创作实践—湘西州武陵山文化产业园规划与建筑设计》
　　　　2016 年 12 月《建筑与文化》
论文 8：《中英住宅性能体系评价比较思路》周湘华、陈克志
　　　　2017 年 6 月《住区》
论文 9：《湖南勘察设计》
　　　　2017 年湖南勘察设计协会专访《寻找初心，一切只是刚刚开始》
论文 10：《基于 BIM 技术的装配式建筑全生命周期的实践与应用》
　　　　2022 年 3 月《工程建筑与设计》周湘华、张国栋

参编规范：
规范 1：:2010 年参编国标《住宅设计规范》GB50096—2011
规范 2：2017 年 8 月参编《湖南省民用建筑信息模型设计基础标准》J 13966—2017
规范 3：2017 年 12 月参编《湖南省建筑工程信息模型交付标准》J 14083—2017
规范 4：2018 年 12 月参编《建筑工程设计信息模型制图标准》J 2628—2019
规范 5：2019 年 6 月参编《湖南省既有多层住宅建筑增设电梯工程技术规程》J 14770—2019

规范 6: 2020 年 12 月参编《湖南省住宅全装修设计标准》J 15524—2021

规范 7: 2021 年参编《工程勘察设计质量管理小组活动导则》

规范 8: 2021 年 8 月参编《医疗建筑项目全咨管理标准》

图集 1: 2017 年主编《多、高层建筑钢结构节点连接选用图集》湘 2017G102—3

图集 2: 2018 年主编《预制生态护坡图集》湘 2018G301

发明 1: 2020 年 11 月高海拔地区低层房屋冷弯薄壁型钢结构体系

课题 1: 《农村社区绿色住宅建筑设计集成技术方案》2017 年 4 月《长株潭两型社会农村社区建设—技术集成与实践》

课题 2: 《湖南省装配式建筑的发展现状调查研究》荣获 2018 年度"知识力量·参政议政学术团队"行动计划重点项目

课题 3: 《湖南省国土空间发展生态修复规划战略调研》荣获 中共湖南省委统战部 2019 年 度无党派人士参政议政专题调研课题评审一般成果项目

课题 4: 《面向老旧小区改造的适老化评估技术集成及示范应用》
住房和城乡建设部标准定额司 2022 年 2 月~2023 年 6 月

课题 5: 《绿色村镇》
湖南省城乡绿色发展重点课题 2022 年 3 月~2022 年 11 月

覆土型博物馆建筑研究

PDCA-BIM 设计全过程管理

内观外筑

工作室历程
The history of the studio

2016.11 湖南 BIM 产业链发展联盟成立大会

2016.12 省建筑师学会年会主讲 BIM 建筑设计优化

2017.1 与影力波结为 BIM 产业链战略合作伙伴

2017.8 赴安化县参加对接易地扶贫搬迁项目建设

2017.12 带领团队赴西藏玉麦乡实地踏勘

2018.4 与湖南城建职业学院产学研讨签约

2019.3 主持公司第一届科学技术委员会会议

2019 年为湖南省注册建筑师继续教育培训授课

2020.8 重庆腾讯双创社区项目督导慰问

2020.9 张家界市城市建设与 BIM 技术应用论坛现场

工作室简介
Studio Introduction

湖南省建科院创作设计工作室，成立于 2010 年，是湖南省建筑科学研究院有限责任公司建筑一院的主力团队，从成立之初的 7 人发展到至今 52 人，团队重点布局城市设计、文化类项目与 BIM 技术 的研究与运用，大力发展文旅、康养、酒店、医疗建筑、特色小镇、教育建筑等新兴领域特色业务。 其团队成员合作创作的湘西武陵山非物质文化遗产展览综合大楼、石门县市民之家、大型智慧社区腾讯双创装配式建筑、 湖南创意总部大厦、韶山创新成果专题展示馆等项目先后荣获全国和湖南省各类 BIM 等大赛的重要奖项。

湖南省建科院业务涉及建筑设计、风景园林、市政工程、水利工程、农业工程、规划设计、装饰设计、工程检测、 BIM 技术应用、工程加固和全过程咨询与管理、技术咨询及培训、建筑行业科技成果评价等。

愿　　景：设计是一场心灵的修行。

价 值 观：科技与文化引领，创意提升价值。

使　　命：将传统文化与当下实践相结合，以服务创造价值。

目　　标：打造学习型团队，建筑幸福人居环境。

行动纲领：勤思考、善总结、敢行动，知行合一。

工作室成员

主 持 建筑师：周湘华

团队管理成员：肖经龙　陈克志　张　平　朱双红　杨　菡　甘海华　李海洲　刘晓光　刘　丹
　　　　　　　陈　涛　周宏宇

专 家 顾 问：戴勇军　黄　严　谭正炎　张云香　陈乐吾　帅东笙　曾益海　金长安　郭　军

成员（排名不分先后）

BIM 组：姜永义　彭　驰　马　胜　卿　楠　曹劲凌　聂　强　许灵波

规划组：王　阳　周　华　汪伟强

建筑组：王国智　朱　海　黄　杰　李　达　周生鹏　肖　威　林业达　晏阳健　文嘉旻
　　　　阳雨涵　陆　枫　于诗琴　孙　文

结构组：郑培明　于志春　杨　石　吴亚玲　彭林茂

设备组：熊海涛　陈永志　杨　晟　张志勇

装饰组：钟　威　邓　皓　崔　琴　刘　娟　章文迪　姚自友　刘佳玲

园林组：柳斯旸

亮化组：唐　昭

研究组：阳　凯　黄晓杰　徐征世

编辑组：易锦田　陈芷芳

后勤组：费　杉　任　泽　戴锐敏　马伊洁　肖　辉　欧阳丽媛

摄影组：何文滔

前期参与：余　鹏　陈　敏　李强国　马颖良　易　凯　郭东亮　苗　杰　欧鹏飞　苏东戈
　　　　　汤林声　郭宏华　臧子禹　张琢瑶　张可喆　桂　琳　夏文辉　李　义　黄龙成
　　　　　贾文沛　刘小军　郭宏华　熊　超　汤林声　李敬良　谢龙龙　李　军

照片墙
Photo Wall

致谢
Thanks

    天道如弓，几许春秋，自 1992 年毕业踏足这个行业，到 2022 年《内观外筑》的出版发行，整整三十年的从业生涯。1995 年参与湘江风光带河西中轴线的城市设计，1999 年在昆明世博会进行现场服务，2003 年完成湘西龙山里耶博物馆设计，2008~2009 年参加清华大学首届建筑设计高级研修班。再到后来的湘西武陵山文化产业园项目设计，石门县市民之家片区和马栏山湖南创意设计总部大厦的设计。一路走来，不断在总结与反思，不断在探索与实践！

    这三十年，大致以十年为一个台阶，分别是起步、提升和实践。头十年，每年的建筑设计项目不多；第二个十年，项目逐渐增多，多以住宅类为主；第三个十年，通过参与诸多的文化类项目，设计也由最初的模仿、借鉴到逐渐形成自己的思考方式，对城市设计、建筑创作以及乡村建筑的绿色化、人居环境适宜化建设也有了不少体会。伴随改革开放，中国建筑设计行业迎来了光辉的历程，设计师的实践机会增多了，但在创作过程中深感自己还需不断学习与反思，做到知行合一。下一个阶段，我将围绕城乡绿色环境发展和提升为目标，以及城市更新和乡村振兴两个子目标，加强对智能建造及数字化转型的研究，通过一系列文旅、康养、建筑工业化的项目来展开实践工作。

    书稿终告断落，掩卷思量，饮水思源，在此谨表达自身的殷切期许与拳拳谢意。建筑学本是一门具有庞大内容结构与实践导向的应用学科，与所有创新成果一样，这要求创作者具有较强的功底与整合能力。在著书过程中作者深刻感觉"学无止境"与"力有不逮"的压力。在此感谢湖南建投相关领导以及导师魏春雨和谭正炎先生，一直以来对我们工作室的指导。2013 年西班牙 FUTURE 杂志赵磊先生，特别邀请 Mangod 大师来长沙讲学，并参与武陵山文化产业园项目方案的设计工作。2015 年与洪达仁先生一起合作了长沙县万绿世界、中山保利文化艺术中心、韶关保利中国武侠城项目，就文旅类设计项目有了不少收获。此外还要感谢张云香、陈乐吾、帅东笙、金长安、影力波李锋（正念先生）和湖南省大众语言艺术研究会演讲专家蒋维等诸位先生对工作室同仁的长期指导。这本书的起源也是在 2015 年与湖南大学王蔚博士和凡益工作室易锦田的交流后，才产生了整理、归纳、回顾的计划。本书在整理修改过程中也得到了阳凯、黄晓杰等研究生的大力协助，2022 年 3 月疫情隔离期间，黄媛媛博士的线上访谈也促进了许多草图的完善与本书的成形。最后要感谢工作室所有工作人员及合作伙伴的大力支持，本书所用之项目均来自于湖南省建科院创作工作室，凝聚了所有参与者的智慧与创新精华，更感谢使本书得以付梓的那些幕后英雄。本书的一些观点和内容，如有不妥之处，还请各位批评指正！

    设计是一场不断修行的过程，建筑师通过项目实践与自己的内心连接，并外化于形，最终指向未来。《内观外筑》只是小小的总结和停顿，非常感谢一路走来的同伴们！

周湘华
2022.10

图书在版编目（CIP）数据

内观外筑 / 周湘华著 . — 北京：中国建筑工业出
版社，2022.10（2023.12重印）
ISBN 978-7-112-27782-7

I. ①内… II. ①周… III. ①建筑设计—研究 IV.
① TU2

中国版本图书馆 CIP 数据核字（2022）第 153348 号

责任编辑：边　琨
书籍设计：易锦田
责任校对：王　烨

内观外筑
Inside Observation Outside Architecture

周湘华　著

*

中国建筑工业出版社出版、发行（北京海淀三里河路 9 号）

各地新华书店、建筑书店经销

北京富诚彩色印刷有限公司印刷

*

开本：787 毫米 ×1092 毫米 1/16　印张：23½　插页：1　字数：458 千字

2022 年 10 月第一版　　2023 年 12 月第二次印刷

定价：269.00 元

ISBN 978-7-112-27782-7

（39961）